Agriculture Extension and Rural Development

NIPA® GENX ELECTRONIC RESOURCES & SOLUTIONS P. LTD.
New Delhi-110 034

About the Editor

Prof. Kalyan Ghadei Ph.D(BHU), B.Mus (ABGM-M),C.C(PI) has a good academic career throughout . Prof. Ghadei has successfully recorded activities at All India Radio, Cuttack, CIFA(ICAR), Bhubaneswar and in OUAT Odisha for transfer of Agricultural, aquaculture and pearl culture technology in Odisha and India.

He is serving as a Professor & Head in the Department of Extension Education in Banaras Hindu University, BHU, Varanasi. He has published more than 60 research papers , authored 09 books and about 25 book chapters. He is a member of many national and international professional societies. Prof. Ghadei has received many awards of national and international repute. The most significant award to his credit is Raman fellowship as a visiting scholar to USA and Post Doctoral fellow at Virginia Tech , USA to study climate change issues and its mitigation strategies.

Agriculture Extension and Rural Development
Dynamics of Sustainable Rural Development and Progress

Editors

Kalyan Ghadei

Associate editors

Musani Chennamadhava
Jagriti Kumari
Harshit Paliwal
Bincy Babu

NIPA® GENX ELECTRONIC RESOURCES & SOLUTIONS P. LTD.
New Delhi-110 034

NIPA. GENX ELECTRONIC
RESOURCES & SOLUTIONS P. LTD.
101,103, Vikas Surya Plaza, CU Block
L.S.C.Market, Pitam Pura, New Delhi-110 034
Ph : +91 11 4386 0225, 9717133558, 9540816132
E-mail: newindiapublishingagency@gmail.com
Website: www. niparesources.com

Print ISBN: 978-93-58879-33-9

ebook ISBN: 978-93-58873-93-1

Composed and Designed by NIPA®.

Dedicated to
All Extension Professionals

Preface

Agricultural extension is a pivotal component in the agricultural landscape, acting as a bridge between research, technology, and farmers. This book aims to unravel the complexities and nuances of agricultural extension, exploring its historical evolution, contemporary challenges, and future prospects.

This book is compilation of extension researches, cases studies from different research stations, KVKs, ICAR organisations and SAUs of India. The nature of article or research may be of different qualities but it is important resource for the policy makers, researchers and academicians. While going through the book the readers will find diverse topics on Agricultural Extension from different agro-climatic zones ranging from crop husbandry to home stead enterprises. The facts, figures, qualitative analysis are very crucial and useful for all agricultural stakeholders.

We acknowledge sincere thanks for the authors who contributed papers and valuable suggestions in making this endeavour a success. Presence of sincere stalwarts of Extension including, Dr. G Trivedi, Dr. V.V. Sadamate, Dr. C. Satapathy, Dr. K. Narayan Gowda, Dr. P. Chandrasekhara, Dr. Manas Mohan Adhikari, Dr. Shankar Acharya, Dr. R. Saravanan and many others are really remarkable. The support of staffs, teachers and students of the Department of Agricultural Extension,BHU Varanasi is appreciable.

The domain knowledge in Agricultural Extension is expanding day by day. We shall be grateful to incorporate comments, suggestions and valuable inputs in future editions

I hope the reading will be informative and satisfying to my extension fraternity.

Kalyan Ghadei

Contents

1

Socio-Economic Profile of Paddy Growers in Milkipur Block of Ayodhya District of Uttar Pradesh

Anurag Dixit, Prakash Singh, R.K. Doharey, Gaurav Kumar, Arvind Kumar and Aman Verma

Department of Extension Education, Acharya Narendra Deva University of Agriculture & Technology, Kumarganj, Ayodhya, Uttar Pradesh-224229

Abstract

The study was conducted in 5 villages of Milkipur block of Ayodhya district which was selected purposively for the study because of the criteria of nearness to researcher and its accessibility. A complete list of all Paddy growers in each selected villages was prepared. From the list. A total number of 120 farmers were selected through proportionate random sampling technique. The results of the study depicted that the majority of the respondents were found in various socio-economic profile characters like, age category of 39-62 years (61.67%), literate (84.16%), general caste (39.17%), nuclear families (85.83%), family size of 4-8 members (62.50%), marginal farmer with a land holding below one ha. (43.33%), Agriculture was observed as main occupation and subsidiary Agriculture labour occupation with (90.83%) and (21.66%), respectively, mix type housing pattern (56.66%), The material possession farm power of farmers' diesel engine (53.33%), chaff cutter (85.83%) was found main farm power and agril. implements, transportation materials bicycle (98.33%), household material coats (100%) and mobile phone (100%) were reported as main communication media possession, (57.50%) respondents had no participation in any organizations. The maximum number of the respondents (52.50%) belonged to annual family income (up to 50000), the maximum number of the respondents Gram pradhan (05.57%),family members (06.00%) and mobile phone (06.00%) was main formal, informal and communicational source of extension contact, respectively.

Keyword: *Socio-economic, Paddy growers.*

Introduction

Rice (*Oryza sativa L.*) is one of the most important cereal grains in the world today and serves as a staple food source for more than half of the world's particularly in India, China and a number of other countries in Africa and Asia. It occupies an area of about 161 million hectares at global level wi`th production of 503.8 million tones milled basis. More than 90% of the rice is produced and consumed in Asian countries. (FAO United Nation 2017).

Rice provides 21 percent of global human per capita energy and 15% of per capita protein. Although rice protein ranks high in nutritional quality among cereals, protein content is modest. Rice also provides minerals, vitamins, and fiber, although all constituents except carbohydrates are reduced by milling.

Rice commodity recognized as a supreme commodity to mankind, because rice is truly life, culture and tradition. It has its own history and religious importance in human life. India is one of the world's largest producers of white rice. For India, like many other developing countries, the issue of feeding ever increasing population is of prime importance, this problem can be solved by maximizing agricultural production through use of high yielding varieties and adopting improved package of practices of rice crop. India is facing the challenges of food and fodder production to meet the demand of rising human and cattle population. One of the major causes of this problem is low level of adoption of improved agricultural practices by the farmers. Population www.thecropsite.com

There are three seasons for growing rice in India viz. autumn, winter and summer. The main rice growing season in the country is kharif. It is known as winter rice as per the harvesting time. The sowing time of kharif rice is June – July and it is harvested in September - October. India is an important centre of rice cultivation. The rice is cultivated on the largest areas in India. Historians believe that while the indica variety of rice was first domesticated in the area covering the foothills of the Eastern Himalayas (i.e. North-Eastern India), stretching through Burma, Thailand, Laos, Vietnam and Southern China, while the japonica variety was domesticated from wild rice in Southern China which was introduced to India. Perennial wild rice still grows in Assam and Nepal.

Among the paddy growing countries, India has the largest area under paddy in the world (45.50 million ha) with a total production of 111.01 million tones during 2017- 18 and it stood next only to China in the world with respect to production India has the world's largest area devoted to rice cultivation. Over half of its rice area is irrigated, contributing 75 per cent of the total production. Notably, this area also consumes 50-60 per cent of the nation's finite freshwater resources. Of the country's 1.15 billion inhabitants, 70% rely on rice for at least a third of their energy requirements. India's population is projected to grow to 1.6 billion in 2050, putting tremendous future strain on its land and water resources India provides around 21 per cent of global rice production from its 28 per cent of the world's rice area. Rice has shaped the culture, diets and economic of thousands of millions of peoples. For more than half of the humanity "rice is life". Considering its importance position, the United Nation designated year 2004 as the "International Year of Rice. Rice is a major food staple and a mainstay for the rural population and their food security. India is an important centre of rice cultivation. Indiastat. com.

Rice is the one of the most important crop in Indo-gangetic plain region of Uttar Pradesh. In India, rice is grown in almost half the states, with West Bengal leading the way in terms of production with 14.71 million tones, followed by Uttar Pradesh (12.22 million tones) and Andhra Pradesh (11.57 million tones) as per the Agricultural Statistics 2014-15, Ministry of Agriculture & Farmers Welfare of the Government of India.

However, the low productivity is a concern for India since India's overall productivity, which lies at 2390 kg/ hectare is less compared to the other countries, as well as there is a stark deference in India's states. Punjab with a productivity of 3,838 kg hectare has the highest productivity in India, and Andhra Pradesh comes second with a productivity of 3,036 kg/ hectare. The divergence shows in the form of Uttar Pradesh which despite being the second largest rice producing state, has a productivity of 2,082 kg/ hectare, much less than the national average. (https://www.geographyandyou.com)

Materials and Methods

The study was conducted in Ayodhya district of Uttar Pradesh it has sufficient grown area under rice crop. That's why it was selected purposively. The Ayodhya district is located in the eastern plain zone of Uttar Pradesh. Out of 11 Community Development blocks in Ayodhya district, which Milkipur block was selected purposively for this study because of the criteria *i.e.* near to Narendra Deva University of Agriculture and Technology, Kumarganj, Ayodhya and its convenient accessibility. five villages were selected randomly in a bolck. Thus the total sample size 120 grown area under rice crop were selected far the present investigation. Data collection designed and exercised by interviewing with few respondents for pre testing. Then the suitable modifications were made according to need of this study. Thereafter the data were collected from the respondents, through personal interview method. Statistical method and analytical tools used for measuring and analyzing the data in the study were-

1. Percentage (%)

The frequency of a particular cell was divided by the total number of respondents or (MPS) in that particular category and multiplied by 100 for calculating the percentage.

2. Average ($\bar{X}$)

The average ($\bar{X}$) was calculated by adding the total scores obtained by the respondents and divided it by the total number of respondents using the following formula:

$$\bar{X} = \frac{\sum X}{N}$$

Where,

$\bar{X}$ = Average or mean

$\sum X$ = Total number of scores obtained by respondents

N = Total number of respondents

3. Standard deviation (σ)

S.D. is the square root of mean of the squares of all deviations, the directions being measured from the arithmetic mean of the distribution. It is commonly developed by symbol (σ).

$$\text{S.D. } (\sigma) = Correl(X,Y) = \frac{\sum (x-\bar{x})(y-\bar{y})}{n}$$

Where,

σ = Standard deviation

d = Deviation of variables mean

M = Total number of items

4. Correlation Coefficient (r)

The coefficient of simple correlation (r) in a measure of the mutual relationship between two variables that in *i.e.* x and y, where relationship is measured and commonly termed as product movement correlation coefficient and is computed by the following formula:

$$Correl(X,Y) = \frac{\sum (x-\bar{x})(y-\bar{y})}{\sqrt{\sum (x-\bar{x})(y-\bar{y})^2}}$$

Where,

X,Y = correlation coefficient

xi = i^{th} value of x variables

$\bar{X}$ = mean of x

Yi = i^{th} value of y variables

$\bar{Y}$ = mean of y

Results and Discussion

1. Age composition

Table 1: Distribution of respondents according to age.

S.No.	Age Categories (years)	Respondents	
		No.	%
1.	Young (up to 38 years)	24	20.00
2.	Medium (39 to 62 years)	74	61.67
3.	Old (above 63 years)	22	18.33
Total:		**120**	**100.00**

Mean = 50.84, S.D. =12.48, Min. =25s, Max. = 80

It is obvious from the Table-1 that highest numbers of respondents (61.67%) were observed in the age group of up to 39 to 62 years, where (20.00%) fall in the age group of up to 38 years and above 63 years of age, there were (18.33%) only.

2. Education

Table 2: Distribution of respondents according to education

S.No.	Categories	Respondents	
		No.	%
A.	Illiterate	19	15.84
B.	Litrate	101	84.16
1.	Primary	18	15.00
2.	Middle	25	20.83
3.	High school	25	20.83
4.	Intermediate	17	14.16
5.	Undergraduate	13	10.84
6.	Post graduate	03	02.50
Total:		120	100.00

The Table-2 focuses that (84.16%) literate and (15.84%) illiterate. Further, the educational level was worked out and given in descending order as (20.83%) middle and high school, (15.00%) primary, (14.16%) intermediate, (10.84%) graduate and (02.50%) post graduate, respectively. It can be concluded that the maximum numbers of respondents were found literate

3. Caste composition

Table 3: Distribution of respondents according to caste.

S.No.	Categories	Respondents	
		No.	%
1.	General caste	47	39.17
2.	Backward caste	36	30.00
3.	Scheduled caste	37	30.83
Total		120	100.00

The Table-3 indicates that the maximum numbers of respondent (39.17%) were belonged to general caste, while the scheduled caste and backward caste were (30.83%) and (30.00%), respectively. Thus, it may be said that the general caste was dominant in the study area.

4. Family type

Table 4: Distribution of respondents according to family type

S. No.	Categories	Respondents	
		No.	%
1.	Nuclear	103	85.83
2.	Joint	17	14.17
	Total:	120	100.00

The Table-4 indicates that (85.83%) respondents were observed residing in nuclear family system as against joint family *i.e.* (14.17%). Hence, it shows that nuclear family system is dominated prevailing in the study area.

5. Family size

Table 5: Distribution of respondents according to size of family

S. No.	Categories	Respondents	
		No.	%
1.	Up to 3 members	30	25.00
2.	4 to 8 members	75	62.50
3.	9 and above	15	12.50
Total		120	100.00

Mean = 05.18, S.D.=2.66, Min.=02.00, Max.=15.00

It is evident from Table-5 that (62.50%) respondent's families were observed such who had 4 to 8 members followed by (25%) families up to 3 members and (15%) respondents families were found having 9 and above members in their families. The average member of family was observed to be 02.66. The range between minimum and maximum number of family members was recorded from 2 to 15. More number of family members might be due to dominancy of nuclear family system in the area.

6. Land holding

Table 6: Distribution of respondents according to land holding

S. No.	Categories	Respondents	
		No.	%
1.	Marginal (below 1.0 ha)	52	43.33
2.	Small (1.0 to 2.0 ha)	27	22.50
3.	Medium (2.0 to 4.0 ha)	25	20.83
4.	Large (4.0 & above)	16	13.33
Total		120	100.00

Mean =01.57, Min.= 0.30, Max.= 6.00

The Table-6 indicates that most of the respondents (22.50%) were found in the land holding category as small (1-2 ha.), (43.33%) in the category of marginal (less than 1 ha.), (20.83%) in the category of medium (2-3 ha.), and in the category of large (Above 3 ha) (13.33%), respectively. The average land holding of the respondents were found to be 01.57 hectare. Thus, it may be said that marginal farmers are more than others in the study area. It seems because of fragmentation of land as a result of deviation of families.

7. Family occupation

Table 7: Distribution of respondents according to family occupation

S. No.	Categories	Respondents			
		Main occupation		Subsidiary occupation	
		No.	%	No.	%
1.	Agriculture labor	01	0.83	26	21.66
2.	Caste based occupation	01	0.83	02	01.66
3.	Service	05	04.16	0	0
4.	Agro-based enterprises	01	0.83	04	03.33
5.	Agriculture	109	90.83	11	09.16
6.	Business	03	02.50	13	10.83
7.	Other	0	0	02	01.66
Total:		120	100.00	58	48.30

It is clear from the Table 7 that in case of main family occupation, the agriculture was observed as main occupation (90.83%) followed by service (04.16%),business (02.50%), and agriculture labor, Caste based occupation, Agro-based enterprises (0.83%). while, in case of subsidiary occupation, the maximum (21.66%) of the respondents engaged in agriculture labor followed by business (10.83%), agriculture (09.16%), agro based enterprise (03.33%), caste-based occupation and other (01.66%), respectively. There were (51.66%) respondents who had not responded on subsidiary occupation.

8. Housing pattern

Table 8: Distribution of respondents according to housing pattern

S. No.	Categories	No.	Percentage
1.	Kaccha	07	05.83
2.	Mixed	68	56.66
3.	Pucca	45	37.50
	Total	120	100

The Table- 8 revealed that majority of farmers have possessed mixed type of house (56.66%) followed by pucca (37.50%), kaccha (05.83%), respectively.

9. Materials Possession

Table 9: Distribution of the respondents on the basis of farm power

S. No.	Categories	Respondents	
		Number	Percentage
1.	Tractor	13	10.83
2.	Power tiller	03	02.50
3.	Diesel engine	64	53.33
4.	Electric motor	18	15.00
5.	Bullock	09	07.50
6.	Tube well/Pumping set	46	38.33

The Table- 9 revealed that the majority of farmers (53.33%) was found having diesel engine followed by tube well (38.33%), electric motor (15.00%), tractor (10.83%), bullock (07.50%) and power tiller (02.50%), respectively.

10. Agriculture Implements Materials:

Table 10: Distribution of the respondents on the basis of agriculture implements

S. No.	Categories	Respondents	
		Number	Percentage
1.	Combine harvester	01	0.83
2.	Cultivator	13	10.83
3.	Seed drill	04	03.33
4.	Thresher	07	05.83
5.	Rotavator	09	07.50
6.	Winnower	19	15.83
7.	Chaff cutter	103	85.83
8.	Disc plough	11	09.16
9.	Deshi plough	05	04.16
10.	Pata	30	25.00

Note: More than one item have been shown by the respondents. Hence, the total percentage of all the items would be more than 100.

It is clear from the data included there in Table-5.1.10 that a majority of the respondents (85.83%) was reported having chaff cutter, pata (25.00%), winnower (15.83%), deshi plough (24.00%), cultivator (10.83%) disc plough (09.16%), rotavetor (07.50%), thresher (05.83%), deshi plough (04.16%), seed drill (03.33%), and combine harvester (0.83%), respectively.

11. Transportation Material Possession

Table 11: Distribution of the respondents on the basis of transportation materials

S. No.	Categories	Respondents	
		Number	Percentage
1.	Truck	01	0.83
2.	Jeep	01	0.83
3.	Pick up	04	03.33
4.	Tractor Trolley	11	09.16
5.	Bike/Scooter	73	60.83
6.	Bullock cart	07	05.83
7.	Cycle	118	98.33
8.	Tempo	10	08.33
9.	Car	07	05.83

Note: More than one item has been shown by the respondents. Hence, the total percentage of all the items would be more than 100.

The Table-11 indicates that an over-whelming majority of the respondents (98.33%) was found having cycle as a means of transportation followed by motor cycle (60.83%), tractor trolley (09.16%), tempo (08.33%), bullock cart and car (05.83%), pick up (03.33%), jeep and truck (0.83%) each, respectively.

Thus, the interference can be drawn from the above data that cycle was important means of transportation with the respondents.

12. Houses Hold Materials Possession

Table 12: Distribution of the respondents on the basis of household materials

S. No.	Categories	Respondents	
		Number	Percentage
1.	Double bed	06	05.00
2.	Sofa set	14	11.66
3.	Dining Table	14	11.66
4.	Dressing Table	12	10.00
5.	Gas stove/Gas cylinder	91	75.83
6.	Electric press	37	30.83
7.	Washing machine	14	11.66
8.	Pressure cooker	94	78.33
9.	Crockery	60	50.00
10.	Fan	91	75.83
11.	Cooler	33	27.50
12.	Solar light	13	10.83
13.	Heater	39	32.50
14.	Cots	120	100.00
15.	Induction cooker	03	02.50
16.	Sewing machine	13	10.83

Note: More than one item have been shown by the respondents. Hence, the total percentage of all the items would be more than 100.

The Table-12 indicates that all respondents were reported that cots followed by pressure cooker (78.33%) followed by gas stove/gas cylinder, fan (75.83%), crockery (50.00%), heater (32.50%), electric press (30.83%), cooler (27.50%), sofa set, dining table and washing machine (11.66%), solar light and sewing machine (10.83%), dressing table (10.00%), double bed (5.00%) and induction cooker (02.50%), respectively. The condition of house hold materials seem to be good.

13. Communication Media Possession

Table 13: Distribution of the respondents on the basis of communication media possession

S. No.	Categories	Respondents	
		Number	Percentage
1.	T.V.	60	50.00
2.	Radio	99	82.50
3.	Mobile	120	100.00
4.	Agril. Journals	08	06.66
5.	Agricultural Magazines	06	05.00
6.	D.T.H.	42	35.00
7.	Journal Magazines	04	03.33
8.	Agriculture Books	06	05.00
9.	News paper	06	05.00
10.	Internet	33	27.50
11.	VCD player	14	11.66

Note: - More than one item have been shown by the respondents. Hence, the total percentage of all the items would more than 100.

Table-13 revealed that all respondents were reported mobile followed by radio (82.50%), T.V. (50.00%), DTH (35.00%), internet (27.50%), V.C.D (11.66%), agriculture journals (06.66%) agriculture books and agriculture books (05.00%) and journal magazines (03.33%), respectively.

14. Social Participation

Table 14: Distribution of the respondents on the basis of social participation

S.No.	Participation	Respondents	
		Number	Percentage
1.	No Participation	69	57.50
2.	Participation in one organization	25	20.83
3.	Participation in two organization	17	14.16
4.	Participation in more than two organizations or office bearer	09	07.50
	Total	120	100

The Table-14 indicates that the majority of respondents had no participation (57.50%) followed by participation in one organization (20.83%), participation in two organizations (14.16%), and participation in more than two organizations (07.50%), respectively. (57.50%) farmers were such who had no participation in any organizations.

15. Annual Family Income

Table 15: Distribution of respondents according to annual family income

S. No.	Categories (Rs.)	Respondents	
		No.	%
1.	(Up to 50000)	63	52.50
2.	(50001to100000)	49	40.83
3.	(100001and above)	08	06.66
Total		120	100.00

Average = 56962.50, Min. Rs. = 20000, Max. = 640000

It is obvious from the Table-5.1.15 that (52.50%) respondents were from those families whose annual income was found in the category of (up to 50000), followed by other categories (Rs. 50001-100000) *viz.*, (40.83%), (06.66%) up to 100001, respectively. The average income was observed to be Rs. 56962.50 with a range of minimum Rs. 20000 and maximum Rs. 640000 only 4 member comes below poverty line, the BPL list has been pegged at about Rs. 27000/annum. Hence, it may be said that the respondents were having considerably good economic condition.

16. Extent of Contact with Information Sources

Table 16: Distribution of the respondents on the basis of extent of contact with different information sources

S. No.	Categories of information sources	Mean score value	Rank order
A. Formal sources			
1.	B.D.O.	01.06	IX
2.	A.D.O.	01.15	VIII
3.	V.D.O.	02.38	VI
4.	Kisan sahayak	03.55	IV
5.	Gram Pradhan	05.57	I
6.	Cooperative societies	04.11	III
7.	Agril. College/ University	01.80	VII
8.	Mandi samiti	03.44	V
9.	Fertilizer / Seed store	04.98	II
10.	Agril. Scientist	0.35	X
	Average	02.83	
B. Informal sources			
1.	Family members	06.00	I
2.	Neighbor	05.74	II
3.	Friends	06.00	I

4.	Relatives	02.96	IV
5.	Local leaders	01.81	V
6.	Progressive farmers	04.81	III
	Average	04.55	
C. Communication media exposure/Contact			
1.	Radio	05.96	II
2.	T.V.	05.81	III
3.	News paper	04.76	VII
4.	Agril. Books	0.87	XIII
5.	News bulletin	05.63	IV
6.	Farm magazines	0.96	XII
7.	Circular letter	0.21	XV
8.	Poster	04.86	VI
9.	Farmers fair	01.27	X
10.	Mobile	06.00	I
11.	Exhibition	01.16	XI
12.	Internet	05.05	V
13.	Film show	03.90	VIII
14	Demonstration	02.15	IX
15	Folders	0.72	XIV
	Average	03.28	
	Over all mean	03.38	

The data furnished in Table-16 pertains to extent of content of respondents with different information sources as used by them for receiving general information as well as about various practices of commercial flower grower. Information sources were categorized in three categories namely formal sources, informal sources and mass media exposure to find out the extent of contact of respondents. So, for as contact with formal sources was concerned, gram pradhan, fertilizer / seed store, cooperative societies, kisan sahayak, mandi samiti, V.D.O., agril. college/ university, A.D.O., B.D.O., and agril. scientist, got the rank orders I, II, III, IV, V, VI, VII, VIII, IX, X, respectively. The mean of scores for all the formal sources was found as (02.83). As for as contact with informal sources was concerned, family members, friends, neighbor, progressive farmers, relatives and local leaders, got rank order I, II, III, IV,V respectively. The mean of scores for informal information sources was found as (04.55).

Among the mass media exposure mobile, radio, T.V., news bulletins, internet, posters, news paper, film show, demonstration, farmers fair, exhibitions, farm magazines, agril. books, folders and circular letters, got rank order I, II, III, IV, V, VI, VII, VIII, IX,X,XI,XII,XIII,XIV,XV, respectively. The mean of scores for mass media exposure was found as (03.38). Hence, it can be concluded that informal sources of information seemed to be most important as generally utilized by most of the respondents. The formal and mass media information sources were also utilized by the respondents with considerable extent. The overall mean of scores for formal, informal and mass media information sources was found to be (03.38) which may be considered as fair contact with information sources.

Conclusions

The results of the study depicted that the majority of the respondents were found in various socio-economic profile characters like, age category of 39-62 years (61.67%), literate (84.16%), general caste (39.17%), nuclear families (85.83%), family size of 4-8 members (62.50%), marginal farmer with a land holding below one ha. (43.33%), Agriculture was observed as main occupation and subsidiary Agriculture labour occupation with (90.83%) and (21.66%), respectively, mix type housing pattern (56.66%), The material possession farm power of farmers' diesel engine (53.33%), chaff cutter (85.83%) was found main farm power and agril. implements, transportation materials bicycle (98.33%), household material coats (100%) and mobile phone (100%) were reported as main communication media possession, (57.50%) respondents had no participation in any organizations. The maximum number of the respondents (52.50%) belonged to annual family income (up to 50000), the maximum number of the respondents Gram pradhan (05.57%),family members (06.00%) and mobile phone (06.00%) was main formal, informal and communicational source of extension contact, respectively.

Reference

Adewale, J.G., Olaniyi, O.A and Adamou, N.A. (2007). Farmers adoption of improved rice technology in Niamy. World Journal Agricultural Sciences. 3 (4): 530-535.

Arathy, B. (2011). Constraint analysis of rice farmers of Trissur district of Kerala. M.Sc. (Ag.) Thesis. ANGRAU, Hyderabad, India.

Awasthi , D.K. (2004). Study on technological gap and constraints analysis of chick pea production technology in Maudaha block of Hamirpur district (U.P) Unpub. M.Sc. (Ag.) thesis submitted to NDUA&T, Kumarganj, Ayodhya.

Balasubramani, N; Swathilekshmi; P.S. and Chandra Kandan, K. (2005). Astudy on the yield gap analysis in paddy in the Erode district of Tamilnadu. Asian J. Ext. Edu, 24 (1&2):44-52.

Chaudhari, R.P.; Singh P. and Mishra B. (2002) Correlates of information sources utilization pattern (ISUA) of rice growers. Indian Res. J.Ext. Edu. 38 (2): 35-40.

Farida, A; Indira, B and Swathi, S. (2011). Communication behavior of paddy farmers. Journal of Communication Studies. 39 (5):51-58.

Kiran, S and Shenoy, S.S. (2010). Constraints in adoption of system of rice intensification in Warangal district of Andhra Pradesh. Journal Research ANGRAU 38 (1 & 2):77-85.

Kumar, G.A.K. and Washon, M. (2007). Improving rice productivity in Orissa : A case of rice development scheme. Indian J. Ext. Edu. 43 (1&2): 49-55.

Mahesh, P;Bhanuprakash, M and Nirajkumar, S. (2011). Farmers empowerment through participatory on farm trials in rainfed rice ecosystem of Koderma, Jharkhand. Indian Journal Extension Education9 (2):23-29.

Man, N and Sadiya, S.I. (2009). Off-farm employment participation among paddy farmers in the muda agricultural development authority and kemasinsemerak granary areas of Malaysia. Asia Pacific Development Journal. 16 (2): 141-153.

Nirmala, K. (2012). A study on diffusion status and adoption of System of Rice Intensification (SRI) in Mahaboobnagar district of Andhra Pradesh. M. Sc. (Ag.) Thesis. Acharya N.G. Ranga Agricultural University, Hyderabad, India.

Obaiah, M. C. (2004). A study on capacity building of rice growing farmers of Farmers Field Schools (FFS) in Krishna - Godavari zone of Andhra Pradesh. Ph.D. Thesis. Acharya N.G. Ranga Agricultural University, Hyderabad, India.

Painkra, S.K., Dev, C.M and Mandal, B.K. (2010). Information sources of tribal rice growers of Bastar district of Chhattisgarh. Journal CommunicationStudies. 28 (9):135-139.

Prasad, R.C.C. (2002). A study on the impact of On-Farm Extension Demonstrations (OFEDs) in Rice in Nellore district of Andhra Pradesh. M.Sc.(Ag.) Thesis. ANGRAU, Hyderabad, India.

2

Socio-Economic Profile of Onion Growers, Kanpur Nagar of Uttar Pradesh

Arvind Kumar, R.K. Doharey, N.R. Meena, Gaurav Kumar, Anurag Dixit Ajay Pratap Singh and Aman Verma

[1,4,5,6,7]*Department of Extension Education, Acharya Narendra Deva University of Agriculture & Technology, Kumarganj, Ayodhya, U.P.*

Abstract

The study was conducted in 5 villages block in Kakwan of Kanpur Nagar district was selected purposively for the study because of the criteria of nearness to researcher Villages and its accessibility. A complete list of all onion growers in each selected village was prepared from the list. A total number of 120 farmers were selected through proportionate random sampling technique. the majority of the respondents were found in various socio-economic profile characters like, age category of 32-51 years (62.50%), literate (71.66%), other backward caste (50.00%), nuclear families (60.84%), family size of 5-8 members (66.66%), marital status married (95.00%), marginal farmer with a land holding below one ha. (52.50%), Agriculture was observed as main occupation and subsidiary Agriculture labour occupation with (89.16%). Kaccha type housing pattern (43.34%). The maximum respondents (44.17%) belonged to the annual income 1 to 2 lakh,. The material possession farm power of farmers' diesel engine (79.16%), Agriculture. Implements khurpi (84.16%), transportation materials bicycle (88.34%), household material gas cylinder, crockery (100%), coats (99.16%). The number of categories was found in medium level of economic motivation and scientific orientation.

Introduction

Onion (*Allium cepa L.)* is one of the most important commercial vegetable crop cultivated extensively in India and it belongs to the family Alliaceae. It is used both in green and mature stage for salad and spice in a variety of flavoured dishes and soups Onion is liked for its flavour and pungency which is due to the presence of a volatile oil 'allyl propyl disulphide'- an organic compound rich in sulphur. Onion bulb is a rich source of minerals like phosphorus, calcium and carbohydrates. It also contains protein and vitamin C. It is being used in several ways as a fresh, frozen, dehydrated bulbs and green bunching types. Dehydrated onion is in great demand which reduces transport cost and storage losses. Onion has got good medicinal value. It contains several anti-cancer agents which have shown to prevent cancer in animals. The beneficial compound called 'quercetin' present in onion has shown to be powerful antioxidant. Extracts of onion are being used in the prevention of 'atherosclerosis' and 'coronary heart disease'. onion extensive culinary, dietary, therapeutic, trading, income and employment generation value. Onion is commodity of mass consumption and is grown almost all over the country mainly by small and marginal farmers (Shukla and Singh-2018).

A global review of onion area and production revealed that onion is grown in the world in an area of 43.64 lakh hectares with a total production of 873.44 lakh tons and productivity of 21.79 tons/ha. China and India are leading onion growing countries with a total production of 247.00 and 159.40 lakh tons, respectively during the year 2017. India produces around 174 lakh tons of onion annually ranking second in the world. Currently in India the production of onion 31129 million tons with the cultivated of 1914 lakh hectares in 2021-22 observed respectively. Among the different onion growing states Maharashtra tops in area and production. (https://www.districts of india.com).

In Uttar Pradesh the major onion producing districts are Fatehpur, Ghazipur, Jaunpur, Farrukhabad, Kannauj, Ballia, Sonbhadra, Mainpuri, Gonda, Hardoi, and Kanpur Nagar in the Uttar Pradesh. The onion production of Kanpur Nagar district was recorded 10.62 MT in 2015 and 10.85 MT in 2016 with the area of onion cultivation of 0.67 M.ha. in 2016. (https://www.districts of india.com).

Materials and Methods

The study was conducted in the Kanpur Nagar district of Uttar Pradesh. Kanpur Nagar district was selected purposively for the study undertaken. It has four Tehsils, namely, Kanpur Nagar, Bilhaur, Kanpur Sadar, Ghatampur and Narval and 10 Community Development blocks, namely (i) Kakwan, (ii) Bilhaur, (iii) shivrajpur, (iv) Chaubepur, (v) Kalyanpur, vi) Bidhnu, (vii) Ghatampur, (viii) Bhitergaon, (ix) Patara, (x) Sarsaul. Out of 10 Community Development blocks in Kanpur Nagar district, the Kakwan block were purposely selected according to need and availability of onion growers. The Block has been divided into three village development officers circle for carrying out the development activities. The revenue villages was arrange in descending order based on the maximum area and maximum number of onion grower, top 5 revenue villages were select from the one block On the basis of maximum area and production. 24 growers were selected from each selected village randomly with the help of progressive growers and village pradhan, thus the total sample size of 120 growers were selected for the present investigation. A structured schedule for data collection will be designed and pretest by interviewing few respondents or farm families to its validity and modify it structure as pre need of change, if any. Thereafter, the data will be gathered with the help of structure schedule by employing personal interview technique.

Statistical Methods Used

1 Percentage (%)

The frequency of a particular cell was divided by the total number of respondents or (MPS) in that particular category and multiplied by 100 for calculating the percentage.

2 Average ($\bar{X}$)

The average ($\bar{X}$) wascalculated by adding the total scores obtained by the respondents and divided itbythetotal number of respondents using the following formula:

$$(\bar{X}) = \frac{\sum X}{N}$$

Where,

$\bar{X}$ = Average or mean

$\sum X$ = Total number of scores obtained by respondents

N = Total number of respondents

3 Standard deviation (σ)

S.D. is the square root of mean of the squares of all deviations, the directions being measured from the arithmetic mean of the distribution. It is commonly developed by symbol (σ).

$$\text{S.D. } (\sigma) = \frac{\sqrt{\sum d^2}}{n}$$

Where,

σ = Standard deviation

d = Deviation of variables mean

M = Total number of items

4 Correlation Coefficient (r)

The coefficient of simple correlation (r) in a measure of the mutual relationship between two variables that in *i.e.* x and y, where relationship is measured and commonly termed as product movement correlation coefficient and is computed by the following formula:

$$Correl\,(X,Y) = \frac{\sum(x-\bar{x})(y-\bar{y})}{\sqrt{\sum(x-\bar{x})^2\sum(y-\bar{y})^2}}$$

Where,

r = Correlation in coefficient

X = mean of all the observation

xi = observation of the variable

Yi = observation of the variables

$\overline{Y}$ = mean of all the observation

Results and Discussion

1-Age of respondents

Distribution of the respondents on the basis of age (years)

S. No.	Age categories (years)	Respondents	
		Number	Percentage
1.	Young (Up to 31years)	20	16.66
2.	Middle (32-51 years)	75	62.50
3.	Old (Above 52 years)	25	20.83
	Total	120	100.00

Mean = 41.71, S. D. =10.39, Min. =22 Max. =63

Revealed that the majority of the respondents i.e. 62.50% were belonged to middle age (32-51years) group, followed by16.66%were young age (Up to 31 years) and 20.83% belonged to old age (above 52years) group respectively. So, the majority of the onion growers fall in the category of 32-51 years of age group.

2. Education

Distribution of the respondents on the basis of education

S. No.	Education level	Respondents	
		Number	Percentage
1.	Illiterate	34	28.34
2.	Literate	86	71.66
	Total	**120**	**100**
3.	Primary Level	9	07.50
4.	Middle level	26	21.66

5.	High school	24	20.00
6.	Intermediate	12	10.00
7.	Graduate	15	12.50
	Total	86	71.66

Revealed that 71.66 percent of respondents were Literate and 28.34 percent respondents were Illiterate. Further, the educational standard of literate respondents in descending order was found as 7.50%, 21.66%, 20%, 10%, and 12.5%, percent for Primary level, Middle level, Intermediate, High school, and Graduate respectively.

3. Caste

Distribution of the respondents on the basis of caste

S. No.	Caste	Respondents	
		Number	Percentage
1.	Scheduled Caste (SC)/ Scheduled Tribe(ST)	43	35.84
2.	Other Backward Caste (OBC)	60	50.00
3.	General Caste	17	14.16
	Total	120	100.00

Mean =1.78, S. D. = 0.67, Min. = 1, Max. = 3

Thus, it is concluded that majority of onion growers (50.00%) belonged to Other Backward Caste (OBC) in Kakwan block.

4. Type of Family

Distribution of the respondents on the basis of type of family

S. No.	Family type	Respondents	
		Number	Percentage
1.	Nuclear family	73	60.84
2.	Joint family	47	39.16
	Total	**120**	**100.00**

Mean =1.39, S. D. =0.49, Min. =1, Max. =2

Show that majority of respondents 60.84 percent belong to nuclear family of the total sample and 39.16 percent onion growers were found in joint type of family. Hence, the dominance of nuclear/single family system was found inthe area.

5. Size of Family

Distribution of the on the basis of size of family

S. No.	Categories	Respondents	
		Number	Percentage
1.	Small family – up to 4 members	22	18.34
2.	Medium family- 5-8 members	80	66.66
3.	Large family- above 9 members	18	15.00
	Total	120	100.00

Mean=6.42, S. D =2.46, Min. = 2 Max. = 16

That majority of the respondents belong to medium size family. This family size (5-8 members) constituted 66.66 percent of the total sample and followed by 18.34 per cent respondents were from small family size (below 4 members) and 15 percent respondents large families size were found above 9 members in their families.

6. Marital status

Distribution of the respondents on the basis of marital status

S. No.	Marital status	Respondents	
		Number	Percentage
1.	Married	114	95.00
2.	Unmarried	6	05.00
	Total	**120**	**100**

The Table 6 indicates that the majority of the respondents (95%) were found married and 5.00% Unmarried respondents.

7. Land holding size

Distribution of respondents on the basis of size of land holdings (in ha.)

S. No.	Size of land holdings (in ha.)	Respondents	
		Numbers	Percentage
1.	Marginal farmers (<1 ha.)	63	52.50
2.	Small farmers (1.0-2.0)	36	30.00
3.	Medium farmers (2.0-4.0)	19	15.84
4.	Large farmers (Above 4.0 ha.)	2	01.66
	Total	**120**	**100.00**

It is indicates that the majority (52.50%) of the respondents were found in the land holding category of marginal farmers (<1 ha) followed by 30% in the categories of small farmer (1.0-2.0 ha), 15.84% in the category of medium farmers (2-4 ha) and 1.66% in the category of large farmers (above 4.0 ha) respectively.

8. Occupation

Distribution of the respondents on the basis of occupation

S. No.	Particulars	Respondents			
		Main occupation		Subsidiary	
		No.	%	No.	%
1.	Agriculture labour	00	00.00	29	24.16
2.	Caste based occupation	03	02.50	13	10.84
3.	Service	06	05.00	00	00.00
4.	Agriculture	107	89.16	07	05.84
5.	Business	03	02.50	00	00.00
6.	Agro-based enterprises	01	00.84	00	00.00
	Total	120	100.00	49	40.84

Indicated that, the Agriculture was emerged as main occupation (89.16%) followed by service (5%), and business (2.5), Caste based occupation (2.5%), Agri-based enterprises (0.84%) and Agricultural based labour there are no any response of the respondents in main occupation.In case of subsidiary occupation, the maximum (24.16) percent of the respondents agriculture labour followed by caste based occupation (10%), agriculture (5.84.%), Business, Agricultural based enterprise and service there was no any response as a subsidiary occupation respectively.

(9) Material possession

(A) Farm materials

(I) Farm power

Table 9(A) (I): Distribution of the respondents on the basis of farm Power

S. No.	Farm power	Respondents	
		Scores	Percentage
1.	Tractor	32	26.66
2.	Power tiller	10	08.33
3.	Diesel engine	95	79.16
4.	Electric motor	18	15.00
5.	Bullock	8	06.66
6.	Tube well	45	37.50
7.	Solar light	16	13.34

That the reveals majority of the onion growing farmers were found having their farm power mainly diesel engine (79.16%) followed by 37.50% 26.66%, 15.0%, 13.34% 8.0%, 6.66 were tube well, tractor, electric motor, Solar light, power tiller and Bullock respectively.

9(A)(II) Agricultural Implements

Table 9 (A)(II): Distribution of respondents on the basis of agriculture Implements

S. No.	Implements	Respondents	
		Scores	Percentage
1.	Cultivator	30	25.00
2.	Seed drill	11	09.16
3.	Zero till cum machine	19	15.84
4.	Habby seeder	18	15.00
5.	Thresher	25	20.84
6.	Rotavator	18	15.00
7.	Winnower	31	25.84
8.	Chaff cutter	96	80.00
9.	Disc plough	20	16.66
10.	Deshi plough	31	25.84
11	Khurpi	101	84.16
12.	Kudal	91	75.84
13.	Savel	27	22.5
14.	Plastic mulch retriever	1	00.84
15.	Ditcher	50	41.66
16.	Seed bed preparation equipments	50	41.66
17.	Tractor drawn ridger	18	15.00
18.	Drip tape layer with auto breake	1	00.84
19.	Stakepuller	7	05.84
20.	Pata	64	53.34
21.	Fogging machine	35	29.16
22.	Sprayer machine	64	53.34
23.	Shorel	45	37.50

Indicated that majority of the onion growing respondents farmer's (84.16%) were reported that khurpi, followed by chaff cutter, (80.00%), kudal, (75.84%), Pata, Sprayer machine (53.34%), Ditcher, Seed bed preparation equipments (41.66%), Shorel (37.50%), Fogging machine (29.16%), Winnower, Deshi plough (25.84), Cultivator(25), Savel (22.5), Thresher (20.84), Disc plough (16.66), Zero till cum machine (15.84%), Stakepuller (5.84), Habby seeder, Rotavator and Tractor drawn ridger (15%), Seed drill (9.16%), Plastic mulch retriever and Drip tape layer with auto breake, respectively.

9 (B) Transportation Materials

Table 9 (B): Distribution of the respondents on the basis of medium of transportation

S. No.	Medium of Transportation	Respondents	
		Scores	Percent
1.	Jeep	19	15.84
2.	Pick up	12	10.00
3.	Tractor trolley	31	25.84
4.	Bike/Scooter	68	56.66
5.	Bullock cart	08	06.66
6.	Bicycle	106	88.34

The data indicates that majority of the respondents (88.34%) were found having Bicycle as a means of transportation followed by bike (56.66%), tractor trolley (25.84%), Jeep (15.84%), pick up (10%), and Bullock cart (6.66) respectively.

9 (C) Household materials

Table 9(C): Distribution of respondents on the basis of household materials

S. No.	Paticulars	Respondents	
		Scores	Percentage
1.	Double bed	31	25.84
2.	Sofa set	08	06.66
3.	Dining table	10	12.00
4.	Dressing table	07	05.84
5.	Gas cylinder	120	100.00
6.	Electric press	44	36.66
7.	Smokeless stove	18	15.00
8.	Pressure cooker	75	62.50
9.	Crockery	120	100.00
10.	Fan	96	80.00
11.	Cooler	34	28.84
12.	Solar light	16	13.33
13.	Heater	07	05.84
14.	Cots	119	99.16
15.	Patromax	14	11.66

The indicates that majority of the onion growing respondents (100%) were reported that gas-cylinder and crockery fallowed by cots (99.16%), fan (80%), Pressure cooker (62.5%), Electric press (36.66%), Cooler(28.84%), Double bed (25.84%), Smokeless stove (15%), Solar light (13.33%), Dining table (12%), Patromax (11.66%), Sofa set (6.66%), Dressing table and Heater (5.84%) respectively. The condition of house hold materials seems to be good.

9 (D) Communication Media Possession

Table 9(D): Distribution of the respondents on the basis of Communication media possession

S. No.	Particulars	Respondents	
		Scores	Percent
1.	T.V.	37	30.84
2.	Radio	31	28.84
3.	Mobile Phone	105	87.50
4.	Tape recorder	07	5.84
5.	Agricultural journal	11	9.16
6.	Agricultural magazines	16	13.34
7.	D.T.H.	34	28.34
8.	Journal	03	2.50
9.	Agriculture book	12	5.80
10.	Newspaper	25	10
11.	Internet	28	23.33
12.	V.C.D./D.V.D. player	07	5.84
13.	Magazines	15	12.50

Indicated that majority the respondents farmers (87.50%) was observed possessing Mobile phone followed by T.V. (30.84%), Radio, D.T.H. (28.84%), internet (23.33%), Agricultural magazines (13.34%), magazines (12.50%), newspaper (10%), Agricultural journal (9.16%), Tape recorder and V.C.D./D.V.D. player (5.84%), Agriculture book (5.80) and Journal (2.50%) respectively. Thus, it may be concluded that the communication media possession viz. mobile phone, T.V., internet, D.T.H., radio and newspaper were main sources for getting information's and recreation purposes.

10. Housing Pattern

Distribution of the respondents on the basis of housing Pattern

S. No.	Housing pattern	Respondents	
		Numbers	Percentage
1.	Kaccha	12	10.00
2	Pacca	62	51.16
3	Mixed	46	38.34
	Total	**120**	**100.00**

Revealed that majority of respondents of onion growers farmers possessed Pacca type of house (51.16%) followed by mixed type of house only (38.34%) and Kaccha type of house (10.00%) was found respectively.

11. Social participation

Table 11: Distribution of respondents on the basis of social Participation

S. No.	Social participation	Respondents	
		Number	Percentage
1.	No participation	42	35.00
2.	Participation one organization	19	15.83
3.	Participation in two organization	42	35.00
4.	Participation more than two organization	16	13.33
5.	Public leader	1	00.84
	Total	**120**	**100.00**

The indicates that out of 120, the majority of the respondents (35%) had no any participation followed by two organization (35%), participation more than two organizations (13.33%) and participation in one organization (15.83%), participation of Public leader (0.84%) respectively.

12. Family Annual Income

The Data presented that the family annual income of majority of respondents farmers 35.84 was found in the very low category of (Up to Rs. 40000) followed by 34.16 percent low category (Rs. 40001 to 60000), 16.66 percent very high (Rs. 100000 and above) 15 percent medium (Rs. 60001 to 80000) and 3.34 percent high (Rs. Rs. 80001to 100000) respectively.

Distribution of the respondents on the basis of annual family income (in rupees).

S. No.	Annual income (In Rs.)	Respondents	
		Numbers	Percentage
1.	Very low (Up to Rs. 40000)	43	35.84
2.	Low (Rs. 40001 to 60000)	41	34.16
3.	Medium (Rs. 60001 to 80000)	18	15.00
4.	High (Rs. 80001to 100000)	04	03.34
5.	Very high (Rs. 100000 and above)	14	16.66
	Total	120	100.00

It is concluded that the majority of the respondent's farmers (35.84%) belonged to very low category (Rs. Up to Rs. 40000) annual income of family respectively. The average annual income was observed to be for minimum Rs. 24000 and maximum Rs. 650000 respectively.

13. Scientific Orientation

Distribution of respondents on the basis of Scientific Orientation

S. No.	Scientific orientation (scores)	Respondents	
		Number	Percentage
1.	Low (up to 18)	23	19.16
2.	Medium (19-24)	82	68.33
3.	High (Above 25)	15	12.50
	Total	**120**	**100.00**

Mean = 21.15, S.D. = 3.45, Min. = 5, Max. =28,

The reveals that 68.33 per cent of respondents belonged to medium level (19-24) of scientific orientation, followed by 19.16 and 12.50 per cent of had low level (up to 18) and them high (up to 25) of scientific orientation respectively.

14. Economic Motivation

Distribution of the respondents on the basis of Economic Motivation

S. No.	Economic motivation (scores)	Respondents	
		Number	Percentage
1.	Low (up to 19)	19	15.84
2.	Medium (20-24)	79	65.84
3.	High (Above 25)	22	18.33
	Total	**120**	**100.00**

Mean = 22.05, S.D. =2.62, Min. = 16, Max. = 29

The reveals that, 65.84 percent of the respondents belonged to medium level of economic motivation category. Whereas,18.3 per cent respondents belonged to high and 15.84 per cent respondents of low level of economic motivation category respectively.

15. Risk Orientation

It can be observed from reveals that, 40.84 percent of respondents belonged to medium level of risk bearing ability, whereas 35.84 and 23.33 percent of them had high and low risk bearing ability respectively.

Distribution of respondents on the basis of Risk orientation

S. No.	Categories (Scores)	Respondents	
		Number	Percentage
1.	Low (up to 16)	28	23.33
2.	Medium (17-22)	49	40.84
3.	High (Above 23)	43	35.83
	Total	120	100.00

Mean = 19.40, S.D. = 4.03, Min. = 10, Max. = 27

Hence, it can be concluded that majority of the respondent onion growers (40.84%) had medium level of risk bearing ability.

Conclusion

It is concluded that majority of the respondents' belong to middle age group. The maximum of respondents was found literate. The majority of the other backward caste, Single families were more in number than joint families. Respondents belonged to single families. Majority of the medium size family. The majority of the respondents married, land holding category of marginal farmers, medium material possession. That majority of respondents of onion growers farmers possessed Pacca type of house, participation followed by two organization. It is concluded that the majority of the respondent's farmers belonged to very low. Majority of the respondents belonged to medium level of scientific orientation. Reveals that of the respondents belonged to medium level of economic motivation category. Majority of respondents in medium level of Risk orientation.

References

Eunice Cavane (2011). Farmers' attitude and adoption of improved Maize varieties and chemical fertilizers in Mozambique. Indian Res. J. Ext. Edn., 11(1):1-6.

Kumar Ashok,B. (2011). A study on entrepreneurial qualities and adoption behavior of banana growers in Gulbarga district of Karnataka. M. Sc.(Agri.) Thesis, Univ.Agric. Sci., Dharwad, Karnataka (India).

Kumari Sangeeta, Kumar Rajesh, Das, S. N. and Kavita .(2019).Performance of Different Herbicides on Weed Control in Onion (Allium cepa L.) and Its Effect on Economics, Current Journal of Applied Science and Technology,33(2): 1-5.

Lakshmisha, (2000) Impact of cashew demonstration on knowledge and adoption and yield levels of farmers in Dakshina Kannada district. M. Sc. (Agri.) Thesis, Univ. Agric. Sci., Bangalore, Karnataka (India).

Madhu, B. M. (2010). Technological gap in turmeric production practices in Belgaum district. M. Sc. (Agri.) Thesis, Univ.Agric. Sci., Dharwad, Karnataka (India).

Mehri Sedigheh (2015). Influence of Planting Date on Some Morphological Characteristic and Seed Production in Onion (Allium cepa L.) Cultivars, Agric. sci. dev.4(2):19- 21.

Patel, J. B.; Patel, A.C. and Sharma(2011). Factors influencing knowledge level of cotton growers about integrated pest management technology. KarnatakaJ. Agric. Sci., 24(4):464-466.

Shukla and Singh (2018). Study of Socio -Economic Status of Onion Farmers in Nashik District of Maharashtra. International Journal of AgricultureSciences, 10(12): 6425-6427.

Suresh Kumar, (2009) A study on technological gap in recommended soybean cultivation practices. M. Sc. (Agri.) Thesis, Univ. Agric. Sci., Dharwad, Karnataka (India).

Tambade, L.R.; Bhise, V.B. and Singh Lakhan (2019). Enhancing Productivity and Quality of Onion through drum roll planting with drip and INM in Solapur district of Maharashtra Indian Res. J. Ext. Edu.,19 (1):87-88.

3

Impact of Frontline Demonstration on Production and Economics of Pulses in Bilaspur District of Himachal Pradesh

[1]Harbans Lal, [2]Shalvi Thakur, [3]Shivalika Sood, [4]Virender Kumar and [5]Rajesh Kumar Thakur

[1,2,4,5]Department of Agricultural Economics, Extension Education & Rural Sociology, COA, CSKHPKV, Palapmur, Kangra, Himachal Pradesh
[3]Assistant professor, Department of Agriculture Economics & Extension School of Agriculture, Lovely Professional University, Phagwara, Punjab

Abstract

Pulses are unique leguminous crops which maintain and restore soil fertility through biological nitrogen fixation. Pulses can be produced with a minimum use of resources and are mostly cultivated under rainfed conditions. Keeping in view the economic importance of pulses, the present study was proposed to study the role of KVK on promotion programme of pulses and the impact of frontline demonstrations on production and economics of pulses in Himachal Pradesh. A manageable sample of 75 farmers (50 FLDs and 25 non-FLDs farmers) was randomly drawn from the villages covered under FLDs. During 2017-18, 98 FLDs were laid down on farmer's fields which comprised of 50 under blackgram and 48 under chickpea. Under lentil crop no FLD was laid down on farmer's field during 2017-18. The gap between the recommended practices in frontline demonstration and farmers' practice of blackgram and chickpea revealed that farmers generally did not use recommended technologies. The technological gaps with respect to all the cultural practices were lower on FLD farms. The productivity of all the crops in case of non-FLD farmers was found to be lower than the productivity of these crops on FLD farms. The influence of experience, area, education and human labour on production was found to be significant for FLD farmers whereas, in case of non-FLD farmers only influence of area and human labour was significant. The study concluded that pulses contributed significantly to the total income of the farmers, therefore it is, suggested that more number of FLDs on pulses must be laid down in the areas having potential of pulse production.

Keywords: *KVK, Frontline Demonstration, Regression analysis, Pulses and Himachal Pradesh*

Introduction

Pulses are the most important part of our diet, richest and cheap source of proteins, vitamins, minerals and amino acids besides high nutritional value. Hence, the economic value of these crops is very high. Pulses can be produced with a minimum use of resources and are mostly cultivated under rainfed conditions and do not require intensive irrigation (Sood et al. 2018). Keeping in view large benefits of pulses for human health; the United Nations has proclaimed 2016 as the International Year of Pulses. Thus, due attention is required to enhance the production of pulses not only to meet the dietary requirement of proteins but also to raise the awareness about pulses for achieving nutritional, food security and environmental sustainability. The Government of India through Indian Council of Agricultural Research (ICAR) has established a wide network of Krishi Vigyan Kendras (KVKs) in all the rural districts of the country. These KVKs under the aegis of the National Agricultural Research and Education System are the real carriers of frontline technologies and impart knowledge and critical input support for the famers. The frontline demonstration programme (FLDs) in pulses is a unique programme by Ministry of Agriculture, Govt. of India, conducted under close supervision of farm scientists. Main objective of FLDs in pulses is to demonstrate and popularize the improved agro-technology on farmers fields under varied farming situations for effective transfer of generated technology and fill the gap between improved technology and adopted/indigenous technology to enhance pulse productivity and farms gains for sustaining the production systems especially under rainfed farming (Choudhary et al. 2009). The field demonstrations conducted under the close supervision of scientists of the National Agriculture Research System is called front-line demonstrations. The main objective of Frontline Demonstrations is to demonstrate newly released crop production and protection technologies and its management practices in the farmers' field under different agro-climatic regions and farming situations.

In Himachal Pradesh, pulses are grown on an area of 31.38 thousands hectare with an annual production of thousand 63.34 MT. The important pulses grown in the state include gram, moong, urd, lentil, rajmash, moth, pea, cowpea, horsegram and lathyrus. However, the farmers in the state are getting high income from pulse production by following modern techniques and recommended package of practices. The soaring prices of pulses in the recent years have also incentivized the farmers to include pulses in their crop production programme. The Krishi Vigyan Kendra of CSK HPKV, located at Berthin in Bilaspur district has a mandate of promoting pulses. The KVK, Berthin has various programmes and demonstrations on pulses. As a result, the farmers' efforts to produce different pulse crops are increasing, leading to higher production and income. Even with the best efforts, pulse acreage and productivity has been stagnant in Himachal Pradesh (Kumar and Prasher, 2012). The productivity of pulses in the state is quite low compared to national and global acreage, mainly due to their cultivation under rainfed and marginal lands besides poor crop management practices (Choudhary 2013; Choudhary et al. 2009). Besides this, Himachal soils are acidic in reaction with low phosphorous availability, thus adequate supply of phosphorous through fertilizers becomes more critical for pulses which is lacking in the farm practices of resource poor hill farmers generally supplying sub-optimal doses of chemical fertilizers especially phosphorous. Thus, keeping in view the above mentioned facts, the present study was undertaken in Bilaspur district of Himachal Pradesh. It was relevant to take up a study entitled, "Impact of Frontline Demonstration on production and economics of pulses in Himachal Pradesh" with following specific objectives:

Objective

1. To examine the impact of Frontline Demonstrations on production of pulses and
2. To work out the economics of pulses in Bilaspur district of Himachal Pradesh

Methodology

The present study was based on both primary as well as secondary data. The secondary data were collected from different reports of KVK, reports of the Department of Agriculture and Directorate of Economics and Statistics. The primary data were collected through well designed and pre-tested schedule by personal interview method. Simple random sampling technique was employed for the selection of farmers covered under FLDs. A manageable sample of 75 farmers (50 FLDs and 25 non-FLDs farmers) was randomly drawn from the villages covered under FLDs. The cost concepts as suggested by commission for Agricultural Costs and Prices (CACP), Govt. of India, 2004 were used to work out the costs and returns of pulse crops produced by both FLD and Non-FLD farmers on their farms.

The technological gap was computed on the basis of difference between the management practices and input usage on progressive/frontier farm and actual management practices and input usage on an average farm. This was computed by using following algorithm:

$Tg = (Yf - Ya) \times 100$

Yf

where,

Tg = Technological gap (%) in input usage

Yf = Input usage by frontier farmer

Ya = Actual input usage by average farmer

To examine the factors affecting the production of pulses in the study area, production functions were estimated using input-output data from individual sample farmers and the linear form of multiple regression model was employed for analysis and discussion as below:

$Y= b_0 + b_1X_1+b_2X_2 +b_3X_3 + b_4X_4+ U$

where,

Y= Production of pulses (q/ha)

X_1= Experience of farmers in pulses production (Years)

X_2= Area under crop in bigha (12.5 bigha = 1 hectare)

X_3 = Educational status of the head of the family

X_4= Total human labour (Mandays/ha)

b_0= Intercept

U= Random term

Results and Discussion

Frontline Demonstration on Pulses

The number of frontline demonstration laid down in different villages from year 2015-2018 have been given in Table 1. In demonstration fields pulses were grown according to the package of practices. It can be seen from the table that during the year 2015-16 a total of 100 FLDs were laid down on farmers field. The number of FLDs increased to 168 in 2016-17 and in the year 2017-18 the number of FLDs conducted were 98 only. In 2016-17 and 2017-18 the maximum yield of 9.20 q/ha and 11.90 q/ha, respectively, was recorded in chickpea (GPF-2) whereas in 2015-16 stagnant yield of 6.10 q/ha was recorded under all varieties of pulses (blackgram and chickpea).

Table 2 reveals that transfer of improved farm technology under frontline demonstrations (FLDs) in pulses resulted in invariably higher grain yield of pulses under demonstration plots than farmers plot yield, which may be attributed to the adoption of recommended agro-technologies in FLDs during study period. Choudhary et al. (2009) has also reported yield enhancement by the use of recommended agro-technologies in FLDs. The table further reveals that the maximum per cent increase in yield over check was recorded to be 51.10 and 49.20 per cent under chickpea (HPG-17) in 2015-16 and 2016-17, respectively. On the other hand, in 2017-18 themaximum per cent increase in yield over check (48.10%) was recorded in Himachal

Table 1: Frontline demonstrations on pulses

Sr No.	Crop	Variety	No.of demonstration 2015-16			Yield (q/ha)					
						2016-17		2017-18			
			2015-16	2016-17	2017-18	Demonstration	Check	Demonstration	Check	Demonstration	Check
1	Blackgram	Him Mash-1	50	60	50	6.10	4.70	6.40	5.30	10.10	9.20
2	Chickpea	HPG-17	39	50	3	6.10	5.30	8.80	5.90	10.60	8.50
		Himachal Chana-2	4	25	38	6.10	5.10	7.90	5.70	11.20	7.56
		GPF-2	7	33	7	6.10	5.40	9.20	6.50	11.90	8.60
	Total		100	168	98						

Chana-2. It also reveals that lowest per cent increase over check in yield was reported in chickpea (GPF-2) in 2015-16, in 2016-17 and 2017-18 lowest per cent increase was reported in blackgram (Him Mash-1).

Table 2: Per cent increase in yield over check

Sr.No.	Crop	Variety	Per cent increase in yield over check		
			2015-16	2016-17	2017-18
1	Blackgram	Him Mash-1	29.80	20.80	9.80
2	Chickpea	HPG-17	51.10	49.20	24.70
		Himachal Chana-2	19.60	38.60	48.10
		GPF-2	13.00	41.50	38.40

Extension Yield Gaps

Extension gaps are the indicators of lack of awareness for the adoption of improved farm technologies by the farmers. The extension gaps in yield of pulses are given in Table 3. Maximum extension gap of 1.40 q/ha was observed in blackgram (Him Mash-1) during 2016-17, 2.90 q/ha gap was observed in chickpea (HPG-17) and 3.64 q/ha in chickpea (Himachal chana-1) during 2017-18. Maximum extension gap of 1.40 q/ha was observed in blackgram (Him Mash-1) during 2016-17, 2.90 q/ha gap was observed in chickpea (HPG-17) during 2017-18 and 3.64 q/ha in chickpea (Himachal Chana-1) during 2017-18. The higher extension yield gaps indicate that there is still a strong need to aware and motivate the farmers for adoption of improved farm technologies in pulses over existing local practices. Refinement in the local farmers practices for higher adoption of location specific generated farm technology for sustaining crop productivity is another option open for the research scientists (Choudhary et al. 2009).

Table 3: Extension gaps in yield of pulses

Sr.No.	**Crops**	**Variety**	**Extension gap (q/ha)**		
			2015-16	**2016-17**	**2017-18**
1	Blackgram	Him Mash-1	1.40	1.10	0.90
2	Chickpea	HPG-17	0.80	2.90	2.10
		Himachal Chana-2	1.00	2.20	3.64
		GPF-2	0.70	2.70	3.30

Pulses Production Technological Gaps

The gap between the recommended practices in frontline demonstrations and farmers' practices of black gram and chickpea in study area are presented in Table 4, respectively. The perusal of the tables revealed that farmers generally did not use recommended and improved technologies. There was a wide gap in use of improved varieties seed in both the crops due to its non availability. In farmers' practice broadcast method of sowing black gram and chickpea against the recommended line sowing was followed and higher seed rate was used. Farmers did not practice seed treatment with *rhizobium* culture, an important component in increasing the yield and yield attributes of pulse crops. Similar observations in seed treatment gap were also reported by Kumar and Elamathi 2007. Partial gap in time of sowing of black gram was also observed. Farmers had sown the black gram in between June15th to June 30th, compared to recommended time of sowing i.e. June end to July beginning. However, no gap in sowing of chickpea crop was observed. These tables further revealed that farmers did not apply any recommended fertilizer, if applied only urea was given to the crop at the time of sowing. Partial or full gap in adoption of weed control and plant protection measures was observed in farmers' practice over recommended practice in frontline demonstrations. Similar observations for gap in improved technologies and farmers practices were also observed by Burman et al. (2010) in different crops.

Crop Area, Production and Productivity

The quantum of various types of crops raised on the farm highlights the economic prosperity and soundness of the farming as occupation. Therefore, the production of major crops on per farm basis has been worked out and displayed in Table 5. It was noticed from the table that among FLD farmers maize gave the maximum production of 7.12 quintal per farm on an average farm followed by wheat (6.48 q/farm) and vegetables (2.61 q/farm). The per farm production of pulses was recorded to be quite low. The table further reveals that in case of Non-FLD farms production of maize was 2.92 q/farm and that wheat(2.52 q/farm). Among pulse crops in case of FLD farmers the per farm production was highest in case of chickpea (1.99 q) followed by blackgram (1.69 q), lentil (1.54 q), respectively. Whereas, in case of Non-FLD farmers per farm production was maximum in case of blackgram (0.30 q) followed by chickpea (0.27 q), lentil (0.20 q), respectively.

Table 4: Comparison between demonstration package of practices and existing farmer's practices of blackgram and Chickpea

Sr. No.	Crop operations	Blackgram			Chickpea		
		Recommended practices demonstrated	**Farmers' practice**	**Gap**	**Recommended practices demonstrated**	**Farmers' practice**	**Gap**
1	Variety	UG-218 and Him mash-1	Local	Full	HPG-17, Himachal channa-2 and GPF-2	Local	Full
2	Land preparation	Two ploughing	One or two ploughings	Nil	Two ploughing	One or two ploughings	Nil
3	Seed rate	20 kg/ha	22-25 kg/ha	Higher	40 kg/ha	50-52 kg/ha	Higher
4	Seed treatment	Rhizobium culture	Nil	Full	Rhizobium culture	Nil	Full
5	Method of sowing	Line sowing at 30cm row spacing	Broadcasting	Full	Line sowing at 30 cm row spacing	Broadcasting	Full
6	Time of sowing	June end to July beginning	June15th to June 30th	Partial	Mid October	Mid October	Nil
7	Fertilizer dose	20:40:20 kg NPK per ha	No fertilizer or urea only	Full	30:60:30 kg NPK per ha	No fertilizer or urea only	Full
8	Method of fertilizer application	Kera	Broadcast at the time of sowing	Full	Kera	Broadcast at the time of sowing	Full
9	Weed management	Pendimethalin application @ 1.5 l ai/ha	No or one hand weeding	Full	Pendimethalin application @ 1.5 l ai/ha	No or one hand weeding	Full
10	Plant protection	Need based pesticide and fungicide application	No pesticide and fungicide application	Full	Need based pesticide and fungicide application	No pesticide and fungicide application	Full
11	Irrigation	Rain fed	Rain fed	Nil	Rain fed	Rain fed	Nil

Table 5: Area, production and productivity of major crops

Sr.No.	Crops		FLD farmers			Non-FLD farmers	
		Per cent area	Production (q/farm)	Productivity (q/ha)	Per cent area	Production (q/farm)	Productivity (q/ha)
1	Maize	17.00	7.12	29.32	19.10	2.92	17.00
2	Wheat	15.53	6.48	21.92	16.32	2.52	18.75
3	Fodder	4.18	2.09	20.34	5.90	1.10	13.88
4	Vegetables	7.65	2.61	28.63	21.18	2.40	21.92
5	Pulses	55.64			37.50		
i	Blackgram	19.32	1.69	4.22	12.85	0.30	2.44
ii	Chickpea	18.55	1.99	5.24	14.93	0.27	2.54
iii	Lentil	17.77	1.54	4.49	9.72	0.20	2.35

The output per unit of area of a crop is represented through yield rate which generally speaks of the economic importance of that crop. For this, the average yields of different crops in the study area have been worked out and are shown in Table 5. On an average FLD farm productivity was found to be 29.32 q/ha for maize, 28.63 q/ha for vegetables 21.92 q/ha for wheat crop respectively. The productivity of pulses on FLD farms was recorded to be 4.22 q/ha, 5.24 q/ha and 4.49 q/ha for blackgram, chickpea and lentil respectively. The per cent difference on both farms was found to be 42, 14, 32 and 23 for maize, wheat, fodder and vegetables. Whereas in case of pulses per cent difference on both farms was 42, 52 and 48 for blackgram, chickpea and lentil, respectively. The productivity of all the crops in case of Non-FLD farmers was found to be less than the productivity of these crops on FLD farms. The productivity of pulses recorded to be in the range of 2 to 3 q/ha.

Technological Gap in Production of Pulses

Technological gap with respect to management practices in the process of production and existing practices of a particular crop indicates the per cent difference in the recommended and existing practices. The technological gap in respect of utilization of various inputs and performing of various cultural practices in the cultivation of pulses has been worked out and is given in Table 6 and Table 7. An overview of the table reveals that on FLD farms the highest technological gap was observed with respect to method of sowing on FLD farms and on Non-FLD farms technological gap was observed to be highest with respect to seed rate. The seed treatment was another practice where significantly high gap was observed on Non-FLD farms. No gap was observed in FLD farms with respect to the use of improved varieties they used seeds of hybrid varieties whereas the gap was recorded to be high (76%) in opting for hybrid seed varieties in Non-FLD farms. Around 24 per cent of gap was observed in case of seed treatment on FLD farms as against of 72 per cent in Non-FLD farms, respectively. Lack of knowledge and ignorance among farmers may be the major reason for this gap. A significantly high gap to the extent was observed for seed rate, land preparation method of fertilizers application, fertilizers doses on Non-FLD farms. It can, therefore, be concluded from the table that the technological gaps with respect to all the cultural practices were lower on Non-FLD farms, as the FLD farmers were having the deep practical knowledge regarding latest cultural and management practices of pulses production and were found to be comparatively more cautious regarding the raising crops in their farms.

Table 6: Technological gap in management practices of pulses

Sr. No.	Recommended practices	Percent of farmers following recommended practices		Technological gap (%)	
		FLD	Non-FLD	FLD	Non-FLD
1	Improved variety	100	24	Nil	76
2	Land preparation	70	40	30	60
3	Seed rate	84	20	16	80
4	Seed treatment	76	28	24	72
5	Method of sowing	62	36	38	64
6	Time of sowing	90	60	10	40
7	Fertilizer dose	64	40	36	60
8	Method of fertilizer application	86	32	14	68
9	Weed management	78	52	22	48
10	Plant protection	94	60	6	40

Technological gap can be defined as the per cent difference in the recommended and existing practices in the process of crop production. Here an attempt has been made to study the technological gap with respect to different agronomic practices viz., seed rate, farm yard manure, fertilizer application, plant protection measures and irrigation. The technological gaps in pulses cultivation on FLD and Non-FLD farms were computed to know the extent of adoption of technology and deviation from the recommended level.

The technological gap in respect of use of inputs such as seeds, farm yard manure, fertilizers doses and application of plant protection measures for FLD and Non-FLD farms is given in Table 7. It is observed that the technological gap in use of seeds was highest in Non-FLD farms and in FLD farms it was less than the Non-FLD farms. This infers that the FLDfarmers were comparatively closer to the recommended practice with respect to seed rate. Again the technological gaps were positive in both the cases which show that both the farm groups were using lesser seed as compared to the recommended. Similarly, a positive and very high technological gap was observed in case of FYM use on both the farms. The farmers used FYM in different crops on the basis of availability of FYM and the quantity of FYM used in the previous crop. They did not keep the recommendations in mind. In case of the fertilizers, higher technological gaps were recorded in FLD farmers than Non-FLD farmers. So far as the plant protection measures were concerned very less technological gaps were observed with respect to the number of sprays of insecticides and fungicides on FLD farmers under study however significantly high gaps were found on Non-FLD farmers. It can be concluded from the table that all the inputs were below their recommended levels more on Non-FLD farmers which indicated that there lies sufficient scope to enhance yields by increasing the use of these inputs.

Table 7: Technological gap in input use under FLD and Non-FLD farmers

Sr.No.	Particulars	Units	Recommended Existing use/application			FLD farmers					
						Technological gaps (%)					
			Blackgram	**Chickpea**	**Lentil**	**Blackgram**	**Chickpea**	**Lentil**	**Blackgram**	**Chickpea**	**Lentil**
1	Seed	kg/ha	25	50	40	20.00	42.10	32.43	20.00	16.00	20.00
2	FYM	q/ha	50	50	50	33.00	25.79	27.30	34.00	48.00	46.00
3	Fertilizers	kg/ha	140	140	140	99.25	104.74	101.35	29.10	25.00	27.85
4	Fungicides	No.	2	2	2	1.46	1.72	1.32	27.00	14.00	34.00
5	Insecticides	No.	2	2	2	1.41	1.68	1.23	29.50	16.00	38.50
						Non-FLD farmers					
Sr.No.	Particulars	Units	Blackgram	Chickpea	Lentil	Blackgram	Chickpea	Lentil	Blackgram	Chickpea	Lentil
1	Seed	kg/ha	25	50	40	17.66	23.72	26.22	32.00	52.00	35.00
2	FYM	q/ha	50	50	50	20.83	22.00	21.77	58.00	56.00	56.46
3	Fertilizers	kg/ha	140	140	140	88.54	94.58	95.24	38.57	32.44	31.97
4	Fungicides	No.	2	2	2	1.12	1.02	1.21	44.00	49.00	39.50
5	Insecticides	No.	2	2	2	1.23	1.41	1.36	38.50	29.50	32.00

Costs and Returns

The analysis of costs and returns is of vital importance both from the point of view of evolving sound production plans and for the formulation of price policy. The study of crop/ enterprise costs and returns also provide very useful information for improving the farm efficiency. The costs and returns for pulse crops grown by the FLD farmers and Non-FLD farmers were estimated and are presented in Table 8. It can be seen from the table that on FLD farms, the total operational cost per ha (Cost A_2) was estimated to be Rs. 35966, Rs. 31911 and Rs. 30362 for blackgram, chickpea and lentil, respectively. In Non-FLD farms the total operational cost per ha (Cost A_2) was estimated at Rs. 19983 for chickpea to Rs. 21984 for blackgram. The comparison across FLD and Non-FLD types of farms revealed that the total operational cost per ha for all pulses was higher by 46 per cent (lentil) to 64 per cent (blackgram) under FLD farms as compared to Non-FLD farms.

The Cost B_2 per ha on FLD farms ranged between Rs. 40133 (lentil) to Rs. 45808 (blackgram). In case of Non-FLD farms the Cost B_2 ranged between Rs. 28967 (chickpea) to Rs. 30904 (blackgram) and were much higher (36% to 48%) for all pulses in FLD farms than Non-FLD farms. The total cost (Cost C_3) per ha was amounted to Rs. 55504, Rs. 52116 and Rs. 47996 for blackgram, chickpea and lentil crop, respectively. The corresponding amounts for the same crops, under Non-FLD farms were recorded to be Rs. 41144, Rs. 36264 and Rs. 35599 for blackgram, chickpea and lentil, respectively. The similar type of difference as that of Cost A_2 and Cost B_2 was noticed in Cost C of all pulses across FLD and Non-FLD types of farms. The gross returns per ha were estimated to be Rs. 44900, Rs. 52400 and Rs. 54860 for lentil, chickpea and blackgram in FLD farms and found to be higher (about 73 to 106%) than Non-FLD farms. In case of Non-FLDfarms the gross returns were recorded to be Rs. 23500 from lentil to Rs. 31720 for blackgram. The table further portrays that the net returns per ha over operational cost (Cost A_2) were recorded to be ranging between Rs. 14538 in case of lentil to Rs. 20489 (chickpea) in case of FLD farmers as against Rs. 2710 (lentil) to Rs. 9736 (blackgram) in Non-FLD farmers. This shows that net returns over operational cost (Cost A_2) for all pulses were higher by about 2 to 5 per cent in FLD farms than Non-FLD farms.

Table 8: Costs and returns from pulses (Rs./ha)

Sr.No.	Particulars	FLD farmers			Non-FLD farmers		
		Blackgram	**Chickpea**	**Lentil**	**Blackgram**	**Chickpea**	**Lentil**
1	Cost A_1	35966	31911	30362	21984	19983	20790
2	Cost A_2	35966	31911	30362	21984	19983	20790
3	Cost B_1	37308	33253	31633	22404	20467	21113
4	Cost B_2	45808	41753	40133	30904	28967	29613
5	Cost C_1	44958	38878	35133	28904	24467	23863
6	Cost C_2	50458	47378	43633	37404	32967	32363
7	Cost C_3	55504	52116	47996	41144	36264	35599
8	Gross returns	54860	52400	44900	31720	25400	23500
Net returns over							
i	Cost A_1	18894	20489	14538	9736	5417	2710
ii	Cost A_2	18894	20489	14538	9736	5417	2710
iii	Cost B_1	17552	19147	13267	9316	4933	2387
iv	Cost B_2	9052	10647	4767	816	-3567	-6113
v	Cost C_1	9902	13522	9767	2816	933	-363
vi	Cost C_2	4402	5022	1267	-5684	-7567	-8863
vii	Cost C_3	-644	284	-3096	-9424	-10864	-12099
Output-input ratio over							
i	Cost A_2	1.52	1.64	1.47	1.44	1.27	1.13
ii	Cost B_2	1.19	1.25	1.11	1.02	0.87	0.79
iii	Cost C_2	1.08	1.10	1.02	0.85	0.77	0.73
iv	Cost C_3	0.98	1.01	0.93	0.77	0.70	0.66

The table also indicates that net returns per ha over Cost B_2 also were recorded to be positive for all pulses in case of FLD farms, however, in case of Non-FLD farms, these were turned out to be positive for blackgram only. The negative net returns per ha of Rs. (-) 3567 and Rs. (-) 6113 were observed for chickpea and lentil, respectively. The net returns per ha over Cost C_1 and Cost C_2 were also turned out to be positive for all pulses under FLD farmers. However, over Cost C_3 the net returns per ha was turned out to be negative for blackgram Rs. (-) 644 and lentil Rs. (-) 3096 and were positive for chickpea (Rs. 284) only. In case of Non-FLD farms, the net return per ha over Cost C_2 and Cost C_3 were turned out to be negative for all the pulses.

Factors Affecting Production of Pulses

To examine the factors affecting the production of pulses, regression analysis was carried out for both FLD and Non-FLD types of farms with the help of linear and log linear functions. However, statistically linear form was found to be the best fit and appropriately on the basis of economic and statistically criterion and was used to study the effect of explanatory variables such as farmers experience in pulses production, area under crop, education of the respondents and total human labour employed in production of different pulses production of blackgram, chickpea and lentil in the study area. The details of regression analysis for FLD and Non-FLD farmers are presented in Table 9 and Table 10. It can be seen from the table that experience of farmers (X_1), area under crop (X_2) and human labour (X_4) had positive significant effect on production of blackgram, the coefficient education (X_3) also turned out to be significant but had negative effect on production of blackgram. This indicate that there was scope for increasing the production as well as profit from blackgram with the increased use of human labour, area under crop and gaining more experience in the pulses production. In case of chickpea crop also the experience (X_1), education(X_3) and human labour (X_4) exhibited positive significant effect on its production. The area under crop (X_2) was turned out to be non-significant in chickpea. The table further shows that in case of lentil, the area under crop (X_2) and education (X_3) had positive significant effect on its production, however human labour (X_4) employed exhibited negative significant effect on production of lentil, this shows that there was over utilization of human labour in the crop, therefore, it can be concluded that the production of lentil could be increased by putting more area under this crop and employment of surplus labour in other pulse crops. The factors included in the analysis were found to most important as they explained as high as 80 to 83 per cent variation in production of pulses in FLD farmers.

Table 9: Estimated regression coefficients of factors influencing pulses production FLD farms

Sr.No.	Particulars	Regression coefficients	Blackgram	Chickpea	Lentil
1	Intercept	b_0	0.6463	0.4218	0.2209
2	Experience (X_1)	b_1	0.0042*	0.0147**	0.056
			(0.0020)	(0.0073)	(0.0317)
3	Area (X_2)	b_2	0.0010*	0.0012	0.0071*
			(0.0005)	(0.0064)	(0.0034)
4	Education (X_3)	b_3	-0.0041**	0.0035**	0.0567*
			(0.0020)	(0.0015)	(0.0259)
5	Human labour (X_4)	b_4	0.1690*	0.0186*	-0.0130**
			(0.0836)	(0.0089)	(0.006)
	Coefficient of multiple determination	R^2	0.8114	0.8330	0.8030
	F value		12.1056	13.1012	9.5421

Note: Figures in parentheses indicate standard errors
*Significant at 5% level of significance
**Significant at 1% level of significance

In case of Non-FLD farmers, the production function explained about 57 to 79 per cent of variation in production of pulses (Table 10). On an average Non-FLD farm, the area under crop (X_2) and the use of human labour (X_4) significantly affected the production of blackgram and chickpea whereas in case of lentil only the factor human labour (X_4) was turned out to be significantly affecting production. The area under crop exhibited significant positive effect on production of blackgram but it exhibited negative significant effect on the production of chickpea. Similarly, the application of human labour in the production of blackgram was observed to over employment as it exhibited negative significant effect on production of blackgram whereas, in case of chickpea it extented positive significant effect. This indicates that production of chickpea could be increased with the more use of human labour. The table further shows that human labour exhibited negative significant effect on production of lentil. This indicates that there was over utilization of human labour (X_4) by Non-FLD farmers in the production of lentil and its use must be diverted to other competing crops. So it can be concluded that there is scope to increase the production and profit from pulses by rationalizing the use of human labour and area under pulse crop.

Table 10: Estimated regression coefficient of factors influencing pulses production Non-FLD farms

Sr.No.	Particulars	Regression coefficients	Blackgram	Chickpea	Lentil
1	Intercept	b_0	0.5885	-0.0078	2.3678
2	Experience (X_1)	b_1	0.0008	0.0034	-0.0087
			(0.0061)	(0.0030)	(0.0091)
3	Area (X_2)	b_2	0.0183**	-0.0010*	0.0041
			(0.0091)	(0.0004)	(0.0049)
4	Education (X_3)	b_3	-0.0002	-0.0003	0.0004
			(0.0007)	(0.0014)	(0.0012)
5	Human labour (X_4)	b_4	-0.0723**	0.1172**	-0.4845**
			(0.0342)	(0.0535)	(0.2373)
	Coefficient of multiple determination	R^2	0.7919	0.6290	0.5718
	F value		9.3567	11.0178	10.0456

Note: Figures in parentheses indicate standard errors
*Significant at 5% level of significance
**Significant at 1% level of significance

Conclusion

The potential of pulses to help and address the future food security, nutritional and environmental sustainability needs has been acknowledged through UN declaration 2016 as international year of pulses. The findings of the study reveals that pulses contributed significantly more than 20 per cent towards the income of the farmer covered under FLD programme. Therefore, it is required that more number of FLDs on pulses must be laid down in the areas having potential of pulse production. There is an urgent need of narrowing down the technological gaps in pulses production in seed rate, method of sowing and plant protection measures in pulse production by bringing more number of farmers under the ambient trainings under required pulse production technologies. More area under production of pulse must be brought by providing incentives to the farmers such as improved seeds and other extension advisory services on weed/pest control and nutrient management.

Refrences

Burman AR, Singh SK and Singh AK. 2010. Gap in adoption of improved pulse production technologies in Uttar Pradesh. Indian Research Journal of Extension Education 10: 99-104

Choudhary AK, Pooniya V, Dass A and Bana RS. 2009. Improved crop management practices for sustainable pulse production: an Indian perspective. Indian Journal of Agricultural Sciences 85: 747-758

Choudhary AK. 2013. Technological and extension yield gaps in pulse crops in Mandi district of Himachal Pradesh, India. Indian Journal of Soil Conservation 41: 88-97

Kumar A and Elamathi S. 2007. Effect of nitrogen level and rhizobium application methods on yield attributes, yield and economics of blackgram (Vigna Mungo L.). International Journal of Agricultural Sciences 3: 179-180

Kumar S and Prasher D. 2012. An analysis of changing trends of foodgrains in Himachal Pradesh. International Journal of Pharmacy and Life Sciences 3: 1739-1742

Sood S, Singh H and Soumya C. 2018. Temporal changes in economics of pulses and their comparative advantage in Rajasthan. Agricultural Science Digest 38: 241-247.

4

Awareness, Knowledge and Perception of Farmers Towards Doubling Farmers Income

***A. Jason Arockiam*[1] *and Ravi Kumar Theodore*[2]**

[1]*Agricultural Extension, Tamil Nadu Agricultural University, Tamil Nadu*
[2]*Agricultural Extension, Directorate of Planning and Monitoring Tamil Nadu Agricultural University, Tamil Nadu*

Abstract

The study aimed to know the farmers awareness, knowledge and perception on the goal of Doubling Farmers Income (DFI) in Coimbatore district of Tamilnadu. Ex-post facto research was carried out and 60 respondents were selected through simple random sampling. The data was collected using a well-structured interview schedule. The respondent's overall awareness of DFI was low (68.33%).About 50.00 per cent of respondents were aware on the goal of DFI. Among the supporting schemes for DFI, majority of the respondents (90.00%) were aware of Pradhan Mantri Kisan Samman Nidhi-PMKSN. Majority of the respondents (96.67%) had low overall knowledge of DFI. About 6.67 per cent of the respondents had known that the goal for DFI was initiated in 2016. All of the respondents (100.00%) didn't know the targeted year fixed for DFI as 2022. About 10.00 per cent of the respondents had known that improvement in livestock productivity was one of the strategies proposed by the government for DFI followed by resource use efficiency (8.33%) and diversification towards high-value crops (6.67%). About 45.00 per cent of the respondents had low overall perception of DFI. About one-fourth (25.00%) of the respondents had opined that DFI by 2022-2023 was achievable followed by scope for increasing their farm yield (50.00%), scope for increasing their farm income (40.00%), scope for reducing the cost of production (30.00%), interested in cultivating multiple farm crops (66.67%), undertaking food processing activities (48.33%), adding more farm enterprises (80.00%) and willing to continue farming in future (70.00%).More than half of the respondents (55.00%) conveyed that their current farm income had decreased while compared to their farm income in 2016.The major constraints in DFI expressed by the respondents were less relevance of minimum support price (98.33%) followed by increasing input costs (96.67%), market fluctuation (95.00%) and inadequate advisories (78.33%). Regulating MSP for all crops, more trainings on Integrated Farming System and food processing, availability of need-based inputs and advisories were the major suggestions from the respondents for doubling their income.

Keywords: Farmer, Income, Awareness, Knowledge, Perception, Constraints

Introduction

The agricultural sector in India has focussed more on agricultural output and food security. From the year 1965 to 2016, India's food production was multiplied by 3.7 times while the population was multiplied by 2.55 times. More than one-fifth of rural households with self-employment in agriculture as their principal occupation were having income less than the poverty line (NSSO, 2011-12). In the early 1980s, the farm income per cultivator was 34 per cent of the income of non-agricultural workers. In the year 1993-1994, the relative income of farmers was one-fourth of the income of the non-agricultural workers. According to SAS 2012-2013, the average annual income of farm households from the farm as well as non-farm sources was Rs. 77,112 which is 60 per cent from farm activities and 40 per cent from non-farm activities. In April 2016, the Government constituted an Inter-Ministerial Committee to examine issues relating to DFI and recommend strategies to achieve the same. The target was to change the farm to non-farm income ratio from the existing 60:40 in 2015-16 to 70:30 by 2022-23. A seven-point strategy was proposed and several initiatives were taken to increase the income of farmers. The last available estimates on the income of agricultural households are based on the Situation Assessment Survey of Agricultural Households conducted by the National Sample Survey Office (NSSO) during 77th round (January – December 2019). As per the survey results, the average monthly income per agricultural household from all sources was estimated to be. Rs. 10218. Thus, the present study was taken to know the awareness, knowledge and perception of farmers towards doubling farmers income. Understanding the farmers awareness, knowledge and perception would help to develop specific strategies for increasing the farmers income.

Methodology

The district Coimbatore and the block Thondamuthur was purposively selected for the study based on the discussions with Agricultural Officers and Assistant Agricultural officers since the farmers in the region were cultivating a wide variety of crops such as cereals, millets, pulses, oil seeds, fibre crops, plantation crops, fruits, vegetables, flowers, medicinal plants, spices and condiments. Three villages were selected from Thondamuthur block based on the highest area under cultivation of various crops viz., Telungupalayam, Pooluvapatti and Devarayapuram. The selected villages were having a land type of wetland, garden land and dryland respectively. Twenty farmers were selected from each village and a total of 60 farmers were selected by simple random sampling technique. Ex-Post Facto Research Design was followed to know farmers awareness, knowledge and perception. The data was collected using a well-structured interview schedule and appropriate statistical tools were used for the analysis.

Results and Discussion

From Table 1, more than two-thirds (71.67%) of the respondents were old and above forty-five years of age. About 30.00 per cent of the respondents had middle education. The majority of the respondents (91.67%) had agriculture as their primary occupation. More than half of the respondents (51.67%) were large farmers having more than four acres of land. More than one-fourth (30.00%) of the respondents had farming experience of more than forty years. About 31.67 per cent of the respondents had an annual income of up to one lakh rupees. Most of the respondents (61.67%) had a medium level of extension agency contact. More than two-thirds (68.33%) of the respondents had a high level of social participation. About 48.33 per cent of the respondents had a low level of mass media exposure.

From Table 2, About two-thirds (68.33%) of the respondents had a low overall awareness of DFI followed by medium level (31.67%). None of the respondents had a high level of overall awareness of DFI. This finding is due to their education, extension agency contact and mass media exposure as evidenced by correlation figures. Among the respondents, only 50.00 per cent were aware of the goal of doubling farmers income which was initiated by the government. The majority of the respondents (90.00%) were aware of Pradhan Mantri Kisan Samman Nidhi – PMKSN scheme followed by Pradhan Mantri Krishi Sinchai Yojana – PMKSY (86.67%), Soil Health Card – SHC (51.67%), Pradhan Mantri Fasal Bhima Yojana – PMFBY (50.00%), National Livestock Mission – NLM (35.00%), Integrated Horticulture Development Scheme – IHDS (11.67%), Mission for Integrated Development of Horticulture-National Horticulture Mission - MIDH-NHM (8.33%), Dairy Entrepreneurship Development Scheme – DEDS (3.33%), Fodder Development Scheme – FDS (1.67%) and National Mission for Sustainable Agriculture – NMSA (1.67%) which supports the farmers in doubling their income. None of the respondents was aware of the schemes National Food Security Mission - NFSM and National Mission on Agricultural Extension and Technology - NMAET.

The majority of the respondents (96.67%) had a low overall knowledge of DFI followed by medium level (1.67%) and high level (1.67%). This finding is due to their education, experience, extension agency contact and mass media exposure as evidenced by the correlation figures. Only 6.67 per cent of the respondents had known that the government had initiated the step towards doubling farmers income in the year 2016. All of the respondents (100.00%) didn't know that the targeted year for doubling farmers income was 2022. About 10.00 per cent of the respondents had known that improvement in livestock productivity was one of the strategies proposed by the government to double the farmers income followed by resource use efficiency/savings in cost of production (8.33%), diversification towards high-value crops (6.67%), improvement in crop productivity (5.00%) and shift from farm to non-farm occupations (5.00%).

About 45.00 per cent of the respondents had a low overall perception of Doubling Farmers Income followed by high level (28.33%) and medium level (26.67%) overall perception. This finding is due to their extension agency contact and mass media exposure as evidenced by correlation figures. Only one-fourth (25.00%) of the respondents had opined that doubling farmers income by the year 2022-2023 was achievable. Half of the respondents (50.00%) responded that they had scope for increasing their farm yield significantly followed by scope for increasing their farm income (40.00%) and scope for reducing the cost of production (30.00%). Around two-thirds (66.67%) of the respondents were interested in cultivating multiple farm crops followed by undertaking food processing activities (48.33%) and adding more farm enterprises (80.00%). More than two-thirds (70.00%) of the respondents were willing to continue farming in future.

More than half of the respondents (55.00%) conveyed that their current farm income had decreased when compared to their farm income in the base year of 2016. Majority of the respondents (81.67%) had expressed that they had no idea to predict the year when their income gets doubled.

Table 1: Profile Characteristics of the Respondents

S.No.	Variables	Categories	Findings	
			f	%
Independent Variables				
1.	Age	Young (Upto 35)	3	5.00
		Middle (36 to 45)	14	23.33
		Old (Above 45)	43	71.67
2.	Education	Illiterate	4	6.67
		Functionally literate	1	1.67
		Primary	7	11.67
		Middle	18	30.00
		Secondary	0	0.00
		Higher secondary	17	28.33
		Collegiate	13	21.67
3.	Occupation	Agriculture as Main Occupation	55	91.67
		Agriculture as Subsidiary Occupation	5	8.33
4.	Farm Size	Marginal (< 1 ha)	1	1.67
		Small (1 to 2 ha)	9	15.00
		Medium (2 to 4 ha)	19	31.67
		Large (> 4 ha)	31	51.67

S.No.	Variables	Categories	Findings	
			f	%
5.	Farming Experience	Up to 10 years	10	16.67
		11- 20 years	6	10.00
		21- 30 years	12	20.00
		31- 40 years	14	23.33
		Above 40 years	18	30.00
6.	Annual Income	Up to Rs. 1,00,000	19	31.67
		Rs. 1,00,001 - 2,00,000	15	25.00
		Rs. 2,00,001 - 3,00,000	8	13.33
		Rs. 3,00,001 - 5,00,000	12	20.00
		Above Rs. 5,00,000	6	10.00
7.	Extension Agency Contact	Low (8 to 13)	17	28.33
		Medium (14 to 19)	37	61.67
		High (20 to 24)	6	10.00
8.	Social Participation	Low (0 to 3)	0	0.00
		Medium (4 to 5)	19	31.67
		High (6 to 8)	41	68.33
9.	Mass Media Exposure	Low (5 to 7)	29	48.33
		Medium (8)	12	20.00
		High (9 to 10)	19	31.67

Table 2: Awareness, Knowledge and Perception of Respondents towards Doubling Farmers Income

S.No.	Variables	Categories	Findings	
			f	%
Dependent Variables				
1.	Awareness	Low (13 to 17)	41	68.33
		Medium (18 to 22)	19	31.67
		High (23 to 26)	0	0.00
2.	Knowledge	Low (9 to 12)	58	96.67
		Medium (13 to 15)	1	1.67
		High (16 to 18)	1	1.67
3.	Perception	Low (8 to 11)	27	45.00
		Medium (12 to 13)	16	26.67
		High (14 to 16)	17	28.33

Constraints Faced by Farmers in Doubling Their Income

The major constraints in doubling farmers income expressed by the respondents were less relevance of minimum support price (98.33%) followed by increasing input costs (96.67%), less market price/low price discover/market fluctuation (95.00%), inadequate advisories (78.33%), small size of landholding (46.67%) and less irrigation facilities (43.33%). Other constraints mentioned by the respondents were man-animal conflict and poor transport facilities.

Table 3: Relationship of Independent Variables with Awareness, Knowledge and Perception of farmers on Doubling Farmers Income

S.No.	Profile characteristics	'r' Value		
		Awareness	Knowledge	Perception
1.	Age	-0.231	-0.213	-0.112
2.	Education	0.293*	0.304*	0.147
3.	Occupation	0.091	-0.241	0.072
4.	Farm Size	0.192	-0.101	-0.205
5.	Experience	-0.091	-0.372**	-0.048
6.	Annual Income	10.151	0.211	-0.180
7.	Extension Agency Contact	0.447**	0.339**	0.348**
8.	Social Participation	0.071	0.101	0.146
9.	Mass Media Exposure	0.313*	0.262*	0.270*

** Correlation is significant at the 0.01 level

* Correlation is significant at the 0.05 level

From the Table 3, it is known that out of the nine independent variables, three variables viz., education, extension agency contact and mass media exposure had positive and significant relationship with awareness on doubling farmers income. The remaining variables were found to have a non-significant relationship with the dependent variable. The correlation value of extension agency contact was significant at 0.01 per cent level of probability, while the correlation values of education and mass media exposure had 0.05 per cent level of probability. It is observed that the awareness on doubling farmers income of the respondents was the function of education, extension agency contact and mass media exposure. Thus, it is concluded that higher the education, extension agency contact and mass media exposure, higher will be the respondents awareness on the goal of doubling farmers income.

From the nine independent variables, three variables viz., education, extension agency contact and mass media exposure had positive and significant relationship with knowledge on the goal of doubling farmers income while the variable experience had negative and significant relationship. The remaining variables were found to have a non-significant relationship with the dependent variable. The correlation value of education and mass media exposure was significant at 0.01 per cent level of probability while extension agency contact and experience was significant at 0.05 per cent level of probability. It is observed that the knowledge on the goal of doubling farmers income of the respondents was the function of education, extension agency contact, mass media exposure and experience. Thus, it is concluded that higher the education, extension agency contact and mass media exposure, higher will the respondents knowledge on the goal of doubling farmers income while higher the experience, lower will the respondents knowledge on the goal of doubling farmers income. Higher the experience, the respondents may not seek information from extension agents and mass media relying on their knowledge and experience. Thus, they may not seek to know the recent initiatives taken by the government.

Out of the nine independent variables, two variables viz., extension agency contact and mass media exposure had positive and significant relationship with perception on the goal of doubling farmers income. The remaining variables were found to have a non-significant relationship with the dependent variable. The correlation value of extension agency contact was significant at 0.01 per cent level of probability while mass media exposure was significant at 0.05 per cent level of probability. It is observed that the perception on the goal of doubling farmers income was the function extension agency contact and mass media exposure. Thus, it is concluded that higher the extension agency contact and mass media exposure, higher will be the respondents perception on the goal of doubling farmers income.

Conclusion

The awareness, knowledge and perception of the respondents on the goal of doubling farmers income were at low level. The extension agency contact and mass media exposure of the respondents were at medium and low level respectively. Efforts should be taken by the extension workers to provide the required information to the farmers at the right time. Only one farmer producer company was functioning at Pooluvapatti village named VelliangiriUzhavan producer company limited. Promotion of FPOs in other villages or nearby may help the respondents in increasing their income. Private extension services were utilized by the respondents in the study area. Linking the public and private extension work will help the respondents address their problems and reach the end farmer. Reducing the cost of inputs will help the respondents in a high benefit-cost ratio which makes the minimum support price beneficial to them. Regulating MSP for all crops, more trainings on Integrated Farming System and food processing, availability of need-based inputs and advisories were the major suggestions from the respondents for doubling their income.

Reference

Ashok Kumar B and Akkamaha Devi Naik. Strategies for doubling farmers' income. The Pharma Innovation Journal. 2021; 10(6S): 430-436.

Gulati A. Budget 2016: 2022 a long toil for tillers of the soil. The Times of India 2017.

https://www.thehindu.com/business/Economy/farmer-producer-organisations-are-coming-in-a-big-way-in-india-nabard-chairman/article36818339.ece

Nandi, Ravi &Swamikannu, Nedumaran. (2019). Agriculture Extension System in India: A Meta-analysis. Agricultural Science Research Journal. 10. 473-479.

Nirmal Rajalakshmi. Real or nominal? Which farm income will Modi double in 5 year. The Hindu Business Line 2017.

P.V Srikala, Doubling Farmers Income by 2022, International Journal of Management (IJM), Volume 11, Issue 1, January 2020, pp. 9–14

PIB, 18 Sep 2020, Steps for Doubling Farmers Income (Press Release) https://pib.gov.in/PressReleasePage.aspx?PRID=1656148

PIB, 20 Jul 2021, Progress in Doubling Farmer Income (Press release) https://pib.gov.in/PressReleasePage.aspx?PRID=1737287#:~:text=The%20strategy%20adopted%20by%20the,returns%20on%20the%20farmers'%20produce

Ramesh Chand, Doubling Farmers Income, NITI Policy paper No 1/2017, National Institution for Transforming India, GoI, New Delhi.

Ramesh Chand. Doubling farmer's income: strategy and prospects. Presidential address at 76th annual conference 2016, Indian Society of Agricultural Economics, Mumbai 2016, 1-20.

Report of the Committee on Doubling Farmers' Income, Volume XI, November 2017.

5

Doubling Farmers' Income and Its Impact in Assam

[1]P. Saikia, [2]A.K. Bhattacharyya, [3]G.Goswami and [4]Varun K. Mahto

[1,2,3]Assam Agricultural University, Jorhat, India
[4]Director, B.V.V. Tiruldih, Seraikella-Kharsawan, Jharkhand

Abstract

Agriculture is the main occupation in India which engages all categories of farmers but the average size of land holding is gradually decreasing due to land fragmentation, industrialization and urbanization which are forcing farmers to quit farming and look for more viable livelihood options. Therefore, Government has taken initiative in doubling of farmers' income which encompasses several dimensions, from production to post-harvest management. In Assam, farmers' income can be increased through increased adoption of supplementary enterprises and producing higher non-farm income through intense skill development initiatives, in addition to increasing agricultural and livestock output.

Introduction

Agriculture is the main occupation in India which engages nearly 119 million of farmers and 144 million landless labourers (Census 2011). According to the Agriculture Census (2012), about 80 per cent of Indian farmers fall under the small and marginal category. The average size of land holding in India is 1.15 ha which will tend to decrease in coming years due to land fragmentation, industrialization and urbanization which are forcing farmers to quit farming and look for more viable livelihood options (Bhattacharyya *et al.* 2018). Specially small and marginal farmers are exposed to high risk of farming due to climate vagaries, uncertainties in production and market demand, lack of access to support services and markets, limited scope of farm mechanization and the consistent failure of half-hearted government efforts which have been unable to make farming an exciting profession to the old as well as new generations. In Indian context, agriculture has never been a profession for the literate section of the society due to lower production and profitability. NSSO report (2012-13), in alignment with the current scenario, says 40 per cent farmers want to quit agriculture if they get alternatives.

Owing to several biotic and abiotic stresses, the problem of agrarian distress in India has accentuated in recent years. Government of India has taken initiative in its annual budget of 2016-17 and initiate a number of steps to alleviate agrarian distress and improve the situation before the fight for food security faces disastrous challenge from mass migration of Indian farmers to other occupations. The steps taken by the Government of India are improving irrigation efficiency, providing crop insurance, improving market infrastructure and its efficiency, promoting organic farming, restoring soil health and so on.

The pathway for doubling of farmers' income encompasses several dimensions, from production to post-harvest management. These include: bridging yield gap, crop diversification, improvements in total factor productivity and proper management of irrigation (GoI, 2007; Evenson and Rosegrant, 1999; Chand *et al.*, 2011; Birthal*et al.*, 2007) along with the provision of market and institutional support for efficient post-harvest management (Saxena and Chand, 2017; Government of India, 2015).

Why Double Farmers' Income?

Past strategy for development of the agriculture sector in India had focused primarily on raising agricultural output and improving food security. Hence raise of farmers' income and promotion of farmer's welfare remained constant, which is evident from the incidence of poverty among farm households. During 1990s, the emergence of agrarian distress in the country due to low level of absolute income as well as large and deteriorating disparity between income of a farmer and non-agricultural worker. Losses from farming, shocks in farm income and low farm income were the major factors related to sharp increase in the number of farmers suicides during 1995 to 2004. The low and highly fluctuating farm income is causing detrimental effect on the interest of the farming community particularly younger age group to leave farming which causes serious adverse effect on the future of agriculture in the country. It is apparent that income earned by a farmer from agriculture is crucial to address agrarian distress (Chand 2016) and promote farmers welfare. In this background, the goal set to double farmers' income by 2022-23 is central to promote farmers welfare, reduce agrarian distress and bring parity between income of farmers and those working in non-agricultural professions.

The concept and Timeframe

Clarity on the following points is important to assess the possibility of doubling the income of the farmers. The substantive points are:

1. What is the period and targeted year for doubling the farm income;
2. What is to be doubled, is it output, value added or income earned by farmers from agricultural activities;

3. Whether nominal income is to be doubled or real income is to be doubled; and
4. Whether the targeted income includes only income derived from agricultural activities or would it also include income of farmers from other sources.

Sources of Growth in Farmers' Income

Doubling real income of farmers till 2022-23 over the base year of 2015-16, needed annual growth of 10.41 per cent in farmers income. This implies that the on-going and previously achieved rate of growth in farm income has to be sharply accelerated. Therefore, strong measures will be needed to harness all possible sources of growth in farmers' income within as well as outside agriculture sector.

The major sources of growth operating within agriculture sector are

1. Improvement in productivity
2. Resource use efficiency or saving in cost of production
3. Increase in cropping intensity
4. Diversification towards high value crops

The sources outside agriculture include

1. Shifting cultivators from farm to non-farm occupations, and
2. Improvement in terms of trade for farmers or real prices received by farmers.

Strategy for improving Farmers' Income

The sources of growth in output and income can be put in four categories.

1. Development initiatives including infrastructure
2. Technology
3. Policies and
4. Institutional mechanisms

Road map and action plan

The quantitative framework for doubling farmers income has identified seven sources of growth. These are:

1. Increase in productivity of crops
2. Increase in production of livestock
3. Improvement in efficiency of input use (cost saving)
4. Increase in crop intensity
5. Diversification towards high value crops
6. Improved price realization by farmers
7. Shift of cultivators to non-farm jobs

Impact of Doubling Farmers' Income in Assam

Assam is a state in north-eastern India, situated south of the eastern Himalayas along the Brahmaputra and Barak river valleys. Assam covers an area of 78,438 km2 (30,285 sq m). Assam's economy is based on agriculture. It produces more than half of India's tea and receives more rainfall than most parts of India; this rain feeds the Brahmaputra River. With the tropical monsoon climate, Assam is temperate and experiences heavy rainfall and high humidity. Based on the rainfall pattern, terrain and soil characteristics, Assam has been delineated into six agro-climatic zones *viz.* North Bank Plain Zone, Upper Brahmaputra Valley Zone, Central Brahmaputra Valley Zone, Lower Brahmaputra Valley Zone, Barak Valley Zone and Hill Zone. The 25 districts were covered 2953 numbers of farm household under doubling farmers' income programme carried out by different categories of farmers in the selected villages. Before beginning DFI efforts in their areas, the Krishi Vigyan Kendras (KVKs) discovered the gaps during the baseline survey. These KVKs strategized to increase farmers' income through increased adoption of supplementary enterprises and to produce higher non-farm income through intense skill development initiatives, in addition to increasing agricultural and livestock output.

Sector-specific Interventions

Field Crops

- Expansion of area and utilization of Rice-fallow with specific crops such as oilseeds crops like toria*var.* TS-67 and Uttara, mustard *var.* NRCHB 101, pulses crop like lentil *var.*Moitre, lathyrus *var.* Ratan, lentil *var.*Moitre and field pea *var.* Aman.
- Execution of INM in toria, lentil and IPM in rice var. Ranjit.
- Technology showcasing and adoption of sali rice *var.* Gitesh, Ranjit sub-1, TTB-404.
- Cultivation of blackgram*var.* PU-31 and short duration paddy variety under post- flood situation.
- Introduction of honey bee in toria cultivation.
- Yield enhancement of matured tea through scientific nutrient management.
- Integrated crop management in buckwheat (GC1), sesamum and toria.

Horticultural Crops

- Cultivation of High value crops such as Assam lemon (local), Banana *var.* Malbhog and Brocolli*var.*Green star.
- Promotion of Sweet Popularization of strawberry *var.*charlie and Early dawn.

- Year round production of cut flowers like summer marigold (Seracole), Gerbera (Red Gem) and Tuberose (Subhasini).
- Scientific cultivation of black pepper *var.*Panniyur 1, potato *var.* Kufri Jyoti and apple ber.
- Popularization of high yielding turmeric *var.* Megha Turmeric 1.

Animal Husbandry

- Introduction of dual purpose improved breeds of Poultry 'Kamrupa', Duck breed 'Indian runner', Dairy, Piggery breed 'Rani, HDK-75 and Goat breed 'Beetel cross breed'.
- Popularization of amur common carp in pond poly culture system.
- Improved feeding and disease management practices.
- Popularization of Integrated fish cum duck farming system.

Farm and non-farm activities

- Use of low cost vermicompost production technology as a resource utilization technology.
- Low cost Oyster mushroom production.
- Scientific bee keeping (*Apis cerana*).
- Production of vegetable seedlings and fruit crop saplings.
- Product diversification and value addition of woven fabric.
- Development of weaving unit.
- Establishment of '*Jalkund*' for life saving irrigation.

Impact on Household Income

The table given below (Table 1) is an analogy between before and after intervention on household income. It can be inferred from the table that the income of the farm households more than doubled from the base year *i.e.* 2016-2017. As far as % sectorial contribution in the total income is concerned, it can be visualized from the table that the share of Field crops and farm and non-farm enterprises sector has been increased over the base year, whereas the share of horticulture crops and fisheries sectors have been reduced during 2020-2021. However the % share of livestock sector in the total income were more or less equal in 2020-21 as compared to base year.

Table 1: Level and change in household income

Crops and Enterprises	Net income (Rs./ household at current prices)		% increase in income	% share in total income		% share in additional income
	2016-17	2020-21		2016-17	2020-21	
Field crops	8952.00	30466.00	240.33	15.61	17.76	13.13
Horticulture	12450.00	35850.00	187.95	21.70	20.90	15.21
Livestock	15680.00	46714.00	197.92	27.34	27.23	38.73
Fisheries	14580.00	39570.00	171.40	25.42	23.06	22.68
Farm and non-farm enterprises	5699.00	18970.00	232.87	9.94	11.06	10.25
Overall	57361.00	171570.00	199.11	100	100	100

Farmers from all land classes benefitted from the technical interventions. In Assam 2953 farmers were surveyed by 25 KVKs. Out of that, it was found that, 43.69% of them were marginal farmers, 33.42% small farmers, 16.97% medium farmers and 5.11% large farmers .However, the households at the bottom of land distribution benefitted relatively more. The income of the landless families increased to 215.75% and marginal and small farm households increased to 210.33% and 207.45% respectively. The medium and large farm households could increase in their household income to 178.75% and 193.37%.

Table 2: Income level and change in household income by land class

Land class	No. of households	% share in total household	Income level (Rs./ household at current prices)		% change in household income
			2016-17	2020-21	
Landless	24	0.81	127863.00	403722.00	215.75
Marginal (<1.0ha)	1290	43.69	36533.00	113373.00	210.33
Small (1-2ha)	987	33.42	54348.00	167098.00	207.45
Medium (2-4ha)	501	16.97	89002.00	248088.00	178.75
Large (>4ha)	151	5.11	138802.00	407210.00	193.37
Total	2953	100.00	446548	1339491	199.97

Conclusion

To fully realise the prospects of agricultural technology, farmer must be linked to extension services and must be guided through their applications, and quality inputs must be delivered on time. Diversification towards high-value crops has the potential to significantly increase farmers' income. Farmers' income can also be increased by expanding into other allied enterprises, such as IFS, rather than relying just on crop production. In all enterprises, appropriate technology use has the ability to increase productivity and farm income. Increase in cropping intensity has the potential to raise farmers' income. Need to find ways to integrate small producers into modern value chains both domestic and export oriented. There is a need to involve state level agencies in creating, nurturing, and establishing FPOs.

Linking processors with producers has vast scope to raise farm income. Promoting food processing in rural areas will also generate employment. Encourage and incentivize efficient use of inputs such as appropriate technology, quality seeds, nutrients, water and energy. Credit must be made accessible, need-based and adequate to fulfil specific individual requirements with a focus on small and marginal farmers. Facilitate and encourage banks to support farmers in situations of natural calamity. Provide infrastructure to reduce wastage and thus improve earnings for farmers.

References

Bhattacharyya S, Burman R R, Sharma J P, Padaria R N, Paul S, Singh A. K. Model villages led rural development: A review of conceptual framework and development indicators. Journal of Community Mobilisation and Sustainable Development, 2018; 13(3):513-26.

Birthal P S, Joshi P K, Roy D, Thorat A. Diversification in Indian Agriculture towards high value Crops: The Role of Smallholders. IFPRI Discussion Paper 00727, International Food Policy Research Institute, Washington, D.C., 2007.

Census, Govt. of India. Retrieved from https://www.downtoearth.org.in/news/farmers-have-decreased-farm-labourers-increased-census-report-40940, 2011.

Chand R, Kumar P, Kumar S. Total Factor Productivity and Contribution of Research Investment to Agricultural Growth in India. Policy paper 25. National Centre for Agricultural Economics and Policy Research, New Delhi, 2011.

Evenson, R E, Pray C, Rosegrant, M W. Agricultural Research and Productivity Growth in GoI (Government of India). Agriculture Strategy for Eleventh Plan: Some Critical Issues. Planning Commission, New Delhi, 2007.

Government of India. Volatility in Onion Prices can be managed through Appropriate Mechanisms and Interventions. Press Information Bureau, New Delhi, 2015. India Research Report No.109. International Food Policy Research Institute, Washington, D.C., 1999.

Roy P, Bhattacharyya S. Doubling farmers' income: Its necessity and possibilities in Indian context. Indian Journal of Agricultural Sciences, 2020; 90 (9): 1639-45.

Saxena R, Naveen P, Singh N P, Balaji S J, Ahuja U, Kumar R, Joshi D. Doubling Farmers' Income in India by 2022-23:Sources of Growth and Approaches. Agricultural Economics Research Review, 2017; Vol.30(2):265-277.

Sexena R, Chand R. Understanding Recurring Onion Price Crisis: Revelations from Production-Trade-Price Linkages. Policy Paper. ICAR-National Institute of Agricultural Economics and Policy Research (NIAP), New Delhi, 2017.

6

Doubling Farmers' Income Through Allied Farm Enterprises: Need to Strengthen Extension Services

Rajesh Kumar Thakur*, Virender Kumar$, Harbans Lal, Kanika Mehta and Komal Sharma

Department of Agricultural Economics, Extension Education & Rural Sociology, College of Agriculture, CSKHPKV, Palampur, Himachal Pradesh

Abstract

Agriculture remains the way of life of majority of rural population in the country. It is the major source of income and employment for rural people. Due to small sized land holdings and lack of concomitant infrastructure facilities, it is difficult to double farmers' income through the traditional farm enterprises in a hilly state like Himachal Pradesh. Hence, there is a need to popularize the allied farm enterprises such as mushroom production, backyard poultry, dairy farming, etc. which are low land demanding and fetching higher returns. The average size of holding in the state has declined from 1.21 hectares in 1990-91 to 0.99 hectares in 2015-16. On an average, this size of holding has a limited potential of generating around 200 mandays and an annual income of Rs 30,000 to 40,000 from the cereal-based cropping systems. Contrarily, economic analysis of allied farm enterprises revealed that round the year production of mushroom (three crops of button, dhingri, milky mushroom of 200 bags unit each) has a potential of generating profit of Rs 70000-80000 /annum. The income from vegetable cultivation under polyhouse conditions varies from Rs 44000 (tomato) to Rs 91000 (capsicum). Similarly, the annual net returns from the small (10 cattle) dairy unit and backyard poultry unit of 25 birds are to the tune of Rs 239000 and Rs18000, respectively. Therefore, there is a need to promote these activities through effective extension services among the hill peasantry so as to harness the complementary/ supplementary enterprise relationships for increasing the farm incomes.

Keywords: *Farmers' Income, Allied Enterprises, Extension Services*

Introduction

Owing to several biotic and abiotic stresses, the problem of agrarian distress in India has accentuated in recent years. The Indian government set a target of doubling farmers' income by 2022–2023 in its annual budget for 2016–17 and started a number of initiatives to achieve this goal, including better irrigation efficiency, crop insurance, market infrastructure improvements, organic farming promotion, soil health restoration, and others. The pathway for doubling of farmers' income encompasses several dimensions, from production to post-harvest management. These include: bridging yield gap, crop diversification, improvements in total factor productivity and proper management of irrigation (GoI, 2007, Evenson et al. 1999) alongwith the provisions of market and institutional support for efficient post-harvest management. With respect to Himachal Pradesh, in order to improve the productivity levels of existing crop enterprises and promote the diversification of cropping systems, various steps have been taken by the line departments which include the creation of water harvesting and irrigation infra-structure, skill up gradation of farmers, subsidy on critical inputs like seed, fertilizers, plant protection materials, market access through information technology and village road network, crop insurance schemes, etc. The impact of these interventions is reflected through the continuous increase in the productivity levels and inclination of farming community towards the adoption of new crop enterprises vegetables and fruit crops.

By and large farming is land based activity and as such the scale of farming enterprises and diversification of farm enterprises is influenced by size of land holding. The farmers with the larger land holdings have relatively more opportunities for increasing the level of income and employment generation. The majority of land holdings in Himachal Pradesh are small and marginal (about 89%), which operate at subsistence level and are not economically viable. It is mainly due to the fact that these are not able to generate full time employment for the farming families. It is not only the case with crop enterprises but also applicable for livestock and other farm enterprises where the benefits of economies of scales are not being harvested due to lack of desirable size of units. When there is a disguised employment, the investment in the infrastructure and machinery is low which may be one of the reasons for the low productivities of farm enterprises. The average size of land holding in the state is continuously shrinking due to the continuous division of holdings due to increase in population and prevalence of nucleus family system. The average size of holding in the state has declined from 1.21 hectares in 1990-91 to around 0.99 hectares in 2015-16. On an average, this size of holding has potential of generating around 200-man days and an annual income of Rs. 30,000 to 40,000 from the cereal based cropping systems. It is asserted that more than half of farm households in the country would remain below poverty level unless they adopt high income earning avenues and augment their incomes through nonfarm activities and high growth in agriculture can reduce income disparities and promote inclusive growth, (Chand et al. 2015).

Hill economy is mainly agro-pastoral and basically rural with nearly 90 per cent of the population living in villages and depending directly or indirectly on agriculture and allied activities (Sarial, 2016). Therefore, there is a need to popularize the agro-based allied farm enterprises which are low land demanding *viz.* mushroom production, backyard poultry, vegetable production, dairy farming, processing and value addition activities etc. This means the farmers' income in Himachal Pradesh can be increased by the few allied farm enterprises taken up along with traditional farm practices. The scale of these enterprises is maintained to such a level so that they may exhibit complementary/supplementary relationship with the existing farm enterprises. On one hand these enterprises would generate continuous flow of cash income and on the other hand income of the farmers will provide regular employment around the year.

Methodology

Himachal Pradesh has varied type of topography and agro-climatic conditions ranging sub-montane, low hills, sub-tropical zone (up 650 meters above mean sea level) to high hills temperate zone (above 1800 above mean sea level). Although, the agro-climatic conditions are congenial for undertaking wide range of agricultural enterprises including vegetables, fruits, flowers, medicinal plants, etc., majority of which are off-season productions for the major markets of the country. The average size of land holding in the state are small hence, the line departments of the state are promoting the agri-enterprises which are less land demanding and has high income and employment generation potential. In this context, protected cultivation of high value crops, mushroom production, small scale dairy farming, backyard poultry have been identified as the key interventions and are being promoted by the state department of agriculture/horticulture/animal husbandry, Krishi Vigyan Kendras (KVKs), State Agricultural Universities (SAUs). For the promotion of these enterprises at the farmers' level various agricultural extension techniques like trainings, demonstrations, exposure visits are being adopted by the stake holders in the different parts of the state. The subject matter specialists of KVKs are in direct contact with the farmers are also laying out Front Line Demonstrations (FLDs) and On Farm Trials (OFTs) and proving time to time consultancies to the farmers. The primary data related to input use pattern, production, price received for produce, etc. of the practicing farmers were collected from the concerned Subject matter specialists of KVKs. The data so collected has been analyzed and pressed through simple tabular methodology.

Results and Discussion

The farmers may adopt one or two aforesaid allied enterprises which fit well to their farm conditions keeping in view the market demand. The potential allied farm enterprises which can be taken up successfully are as under:

1. Mushroom Cultivation

With the increase in socio-economic conditions of the people the as well as awareness about the nutritive value of the mushrooms, their demand is continuously increasing among different sections of the society. The mushroom production has been a common practice among farmers of the state where the winter season is relatively prolonged like Solan, Shimla, etc. It was so because it is the climate which suits for the production of button mushroom. In present scenario, in addition to the button mushroom, *Dhingri* and milky mushrooms which can be grown in warmer climate are also popular among mushroom growers of the state.

Feasibility of Production

Most of the farm households have sufficient residential accommodation from which rooms for the production of mushrooms can be spared.

Climatic Requirement

In the present scenario the mushroom production can be undertaken round the year under the natural climatic conditions. Button mushroom is most commonly grown during the winter season as it requires relatively cold climatic conditions (15-25° C) which are available in the winter months of the year from October to February. *Dhingri* is suitable for warm climatic conditions where the temperature ranges between 25-30° C. The places or the period during which the temperature rises beyond 30° C, milky mushroom can be grown successfully. The overall yield improvement in first and second seeding in the month of January and February is due to optimum temperature and relative humidity for the development of mycelium, which helped in improving the yield of mushroom, Chitra et al. (2018). The temperature requirements and periods under which different categories of mushroom can be grown in the state are indicated in the Table.1

Table 1: Temperature requirements and suitable time periods for growing different categories of mushrooms under natural climatic conditions

Sr. No.	Type of Mushroom	Temperature requirements	Suitable period during the year
1.	Button mushroom	15-25° C	October to February
2.	*Dhingri*	25-30° C	February to April and September to October
3.	Milky Mushroom	30-40° C	May to August

The periods of mushroom production may be varied under the controlled conditions i.e. with the use of air conditioners and blowers. But the cost of production of mushroom will be comparatively higher with the use of air conditioners and blowers. This technology is useful only if the demand of a particular variety of mushroom is high and prices are remunerative, otherwise, it is economical to produce under the normal climatic conditions as suggested above.

Economics of Mushroom Production

In general, during a period of one year, three crops of mushroom can be taken under natural climatic conditions as indicated in Table 1. The economics of a mushroom house for unit of 200 bags has been worked out and is presented in Table 2. The table indicates that an initial investment of Rs. 29600/- is required. Based on this investment the total annual fixed cost on account of interest on fixed investment and depreciation is around Rs. 5920. Since, the mushroom house can be used for the production of three crops during a year, hence the total fixed crop is distributed equally among three crops i.e. Rs. 1973/crop. The working expenditure for production of mushroom includes the purchase of mushroom compost or straw and spawn, chemicals, electricity charges, labour, etc. The total variable cost in case of button mushroom is comparatively higher as compare to the *dhingri* and milky mushroom, Rs. 25,911, Rs. 18767 and Rs. 19,270, respectively. The variable cost of the *dhingri* and milky mushroom is low due to the reason that the compost can be prepared by farmer itself. The total cost of maintaining a production unit of 200 bags (10 kg capacity) varied between Rs. 20,740 in case of *dhingri* to Rs. 27,884 in case of mushroom. The total production of mushroom from the unit is around 600, 500 and 600 kg in case of button, *dhingri* and milky mushroom, respectively and similar results were also observed by Sarial (2016). The net returns of the crops will be highest in case of button mushroom followed by milky mushroom. It can be concluded that the mushroom production may be one of the feasible enterprises for increasing the farm household income by earning a additional income of Rs. 70,000-80,000. The returns from the enterprise may further be increased through process and value addition activities.

Table 2: Economics of Button, *Dhingri* and Milky Mushroom production unit (200 bags)

(Rupees)

Sr. No.	Particulars	Type of Mushroom		
		Button	*Dhingri*	Milky
A.	**Fixed Investment**			
i.	Modification of rooms to mushroom house (10'x20')	10000		
ii.	Construction of racks for placing compost bags	15000		
iii.	Basic tools and utensils			
a)	Spray pump	1200		
b)	Thermometer	700		
c)	Bucket	200		

Sr. No.	Particulars	Type of Mushroom		
		Button	***Dhingri***	**Milky**
d)	Weighing machine	1500		
e)	Miscellaneous tools	1000		
	Total Fixed investment	29600		
B.	Fixed Cost (interest on fixed investment &depreciation @10% each)	1973	1973	1973
C.	Variable cost (Rs./crop)			
i	Mushroom Compost bags (200 bags @ Rs 100/bag)	20000	-	-
ii	Wheat straw (500 kg @Rs 7/kg)		3500	3500
iii	Formalin (1200 ml) and Bavistin (84g) for sterilization of straw	-	800	800
iv	Spawn (40 kg @ 225/kg)	-	9000	9000
v	Polythene bags (90x60 cm) 200 @ Rs 3/bag	-	600	600
vi	Labour (about 15 man days @250/man day)	3750	3750	3750
vii	Casing soil	500	-	500
viii	Miscellaneous expenses like electricity, other chemicals	1500	1000	1000
ix	Interest on working capital @5% for 1.5 months)	161	117	120
	Total Variable cost (Rs)	25911	18767	19270
D.	Total cost of production (B+C)	27884	20740	21243
E.	Production of mushroom (kg)	600	500	600
F.	Average Price (Rs/kg)	100	80	80
G.	Gross Income	60000	40000	48000
H.	Net Income	32116	19260	26757

2. Backyard Poultry

Poultry is one of the important enterprises for the self-employment of educated youths. Poultry can be successfully undertaken as a commercial as well as allied farm activity in rural areas to supplement the household income. Meat and eggs are the major products of the poultry and the demand of these products is continuously increasing due to increase in population and their importance in daily life on account of their nutritive value. These two products of the poultry are the rich source of energy and proteins and are recommended by the dieticians for maintaining proper health of different age groups of people. In the total egg production, the share of unorganized poultry production is around 20 per cent. Organized or commercial poultry production in small areas of local vicinities is not possible on account of lack of desired demand and space. However, backyard poultry units as a supplementary enterprise may be a useful and feasible option to create additional sources of income and employment generation as the enterprise can be run round the year.

Feasibility of the Enterprise

Institutional support

Promotion of backyard poultry in the rural areas is one of activities of the State department of animal husbandry and breeding and is being implemented in different parts of the state. Under this scheme the farmers are provided chicks at subsidized rates and a financial assistance worth Rs. 1200/unit for the establishment of unit. In addition to this, the department of Animal Genetics and Breeding, College of Veterinary & Animal Sciences, CSK HPKV, Palampur is also promoting backyard poultry in Kangra, Chamba and Lahaul &Spiti districts of the state under All India Coordinated Research Project. The department is also providing free chicks and feed to start with the production. In both the schemes, time to time technical support is provided to the farmers for management of birds.

Farmers' Level

The farming in the villages is of subsistence and integrated nature where the crops, livestock and other farmer enterprises are managed in a holistic approach as single unit. In addition to the residential building, the farmers also have cattle shed for the rearing of milch animals. Adjoining to the cattle shed the farmers also keep some open area. The backyard poultry unit can be hosted in small shed either adjoining to the cattle shed or in the open space area. The number of birds is less as compared to the commercial units and they may survive on scavenging and other household wastes. However, if the wastes are not sufficient then the supplementary feeding may also be done. The unit can be maintained for both eggs as well as meat production.

Income from Backyard Poultry

Since, backyard poultry is taken as an allied activity by the rural households exhibiting no competition for other farm enterprises, therefore the number of birds may be kept at around 25 per unit only. These birds can be maintained at zero maintenance cost because their requirements may be met from usual household and farm wastes especially in case of native birds. However, the requirement for maintenance of *Himsamridhi*variety (akin to native birds) developed by COVAS, CSK HPKV is higher as compared to the native birds, in which the supplementary feeding is required for the better performance. The growth and production potential of this variety is significantly better than the native birds. The *Himsamridhi* birds lay 130-140 eggs as compared to the 60-70 in case of native birds. The data collected with respect to production and management costs from the adopted farmers has been analysed and is depicted in Table 3. It can be observed from the table that the total cost of maintaining a backyard poultry unit of 25 birds is around Rs. 1820/- in case of native birds and Rs. 2948/- in case of *Himsamridhi* birds per annum. In the total annual cost of maintaining of a backyard poultry unit the share of fixed cost is low as compared to variable cost i.e. it was around 55 and 34 per cent in case of native and *Himsamridhi* birds, respectively. As far as the annual maintenance cost i.e. variable cost is concerned, it is estimated at Rs 820 and 1948 for native and *Himsamridhi* birds respectively.

The income from the unit may be derived from the sale of eggs and birds for meat purpose. The income from the sale of eggs and birds for meat was estimated at Rs 13500 and Rs 7500, respectively, in case of *Himsamridhi* unit. The table clearly indicates that the income from the *Himsamridhi* unit was comparatively higher as compared to the native birds' unit. The net income from the poultry unit was estimated at Rs. 10,380 and 18,052 per annum for native and *Himsamridhi*unit, respectively. This clearly indicates that by maintaining a backyard poultry unit, a rural household will be able to generate reasonable profits with minimum efforts and resources. Earlier Islam et al. (2015) have also reported that the maximum amount of income was contributed by selling of eggs (40.00 %) followed by sale of cocks (34.36 %) and sale of spent hens (25.64 %) in case of local chicken. Higher B:C ratio of 4.41, 2.62, 2.60 and 5.57 in the improved breeds of backyard poultry was also observed by Singh et al. (2019), Baruah et al. (2018), Islam et al. (2015) and Uddin et al. (2013), respectively. Besides income generation, backyard poultry helps in alleviation of malnutrition of the rural people through the production of valuable animal protein and empowers rural women (Deka et al. 2014).

Table: 3: Economic Assessment of *Himsamridhi* birds as compared to Native/desi birds

Sr.No.	Particulars	Type of bird	
		Native	***Himsamridhi***
1.	Investment on poultry house	5000	5000
2.	Fixed cost		
	Interest on investment on poultry house	500	500
	Depreciation of poultry house	500	500
	Total	1000	1000
3.	**Variable cost**		
	Purchase of chicks @	300	300
	Supplementary feeding etc.	-	1100
	Miscellaneous expenditure	500	500
	Interest on total working capital @5% for half year	20	48
	Total	820	1948
4.	Total Cost (Fixed+ Variable cost)	1820	2948
5.	Production of eggs	650	1350
6.	Sale of eggs @ 10/egg	6500	13500
7.	Sale of male bird	3200	4000
8.	Females	2500	3500
9.	Gross income (6+7+8)	12200	21000
10.	Net profit (9-4)	10380	18052
11.	Net Returns per bird	415	722

3. Small Scale Dairy

Animal husbandry is the most important component of Indian agriculture supporting livelihood of the rural population especially marginal, small and landless farmers (Singh et al. 2018). At constant prices it was observed that the value of output from livestock is about 29% of the value of the output from total agriculture and allied sector (DAHDF, 2016-17). Within Livestock sub-sector, dairying constitutes the major share about 67% in value of outputs from agriculture (Jaiswal et al. 2018). Livestock sector provides employment to 8.8% of population which largely comprises of landless and unskilled population. The demand of milk and milk products is increasing on account of improvement in social status and population. Moreover, the consumption of milk and milk products make the human diet nutritious and balanced especially in case of vegetarian population. In the state, the farming community has a practice of maintaining 1-2 milch animals (cattle/buffaloes) to meet the day to day milk requirement of the family. The existing cattle and buffalo population of the state is not in a position to meet out the total demand for milk and milk products of total population of the state. As a result, huge quantity of processed milk, sweets, *paneer* etc. is coming from the adjoining state. This indicates that there exists a huge potential in the state for promoting the dairy farming activity. The existing practice of dairying in state is of subsistence nature and family oriented. Moreover, maintenance of 1-2 milch animals is not an economically viable option because the resource use efficiency in this case is very low. If the size of unit is increased then the economies of scale may be realized and resource use efficiency will also be high. By increasing the number of animals, the farmers will also be able to produce a sizeable marketable surplus and as well as process the milk for preparation of value-added products like curd, *paneer, ghee, khoya* etc. Here dairy farming plays a significant role in doubling farmers' income and it needs to maintain an annual growth rate of 5.3 % in milk production. Farmers should be aware of recent technologies interventions and focus on keeping healthy animals for producing good quality and quantity of milk (Sahu et al. 2019). So, it can be said that success of a dairy depends upon the factors like; availability of animals of suitable breed, fodder (green and dry), concentrates (feed), technical know-how and availability of market for the disposal of produce. Thus, an attempt has been made to work out the economics of small dairy unit of 10 cattle, which may be manageable at farmers' level. The capital requirements and cost & returns pattern of small dairy unit is discussed as under:

Capital Requirement

Just like other production enterprises, the dairy farm also requires two types of capital i.e. fixed capital and working capital. The fixed capital is the investment which is relatively for longer period of time and is associated with basic infra-structural facilities and assets. This includes investment for the construction of dairy shed, machineries and implements like chaff cutter, tools, implements and utensils etc. On the other hand, working capital are the items which are required for the day to day functioning of the dairy unit. The working capital may include the expenditures on fodder (green and dry), concentrates, mineral, medicines and health care, human labour, marketing expenses, etc. the capital requirements for a dairy unit of 10 cattle have been estimated and depicted in Table 4.

Table 4: Capital requirements for a dairy unit of 10 cattle (Jersey crossbred) (per annum)

Sr.no.	Particulars	Quantity	Rate (Rs)	Amount (Rs)	Per cent to total
A.	**Total Fixed Investment**				
i	Construction/improvement of existing cattle shed	1	500000	500000	55.87
ii	Cattle (Jersey crossbred)				
a	In milk	6	40000	240000	26.82
b	Dry/pregnant	4	30000	120000	13.41
iii	chaff cutter with electric motor	1	20000	20000	2.23
iv	Implements, utensils etc.			15000	1.68
v	insurance	10	1000/cattle	10000	1.12
	Total			895000	100
B.	**Working Capital**				
i	**Grass**				
a	Green (25kg/cattle/day)	913	200/q	182600	22.07
b	Dry (5kg/cattle/day)	183	700/q	128100	15.48
ii	Concentrates (4kg/cattle/day during milking and 3kg/cattle/day in dry period	140	2000/q	280000	33.85
iii	Medicine & health care		500/cattle	5000	0.6
iv	Electricity etc.			3600	0.44
v	Human labour (2 man days/day)		250	180000	21.76
vi	Milk marketing		3000/ month	36000	4.35
vii	Misc. Charges			12000	1.45
	Total			827300	100

It can be observed from the table that the total requirement of the fixed capital will be to the tune of Rs. 8,95,000. Among the different component of fixed capital, the major portion (55%) will be invested for the construction or modification of the cattle shed. The next major item of the fixed investment is the purchase of milch animals which was estimated at Rs. 3,60,000 accounting for about 40 per cent of the total fixed investment. The total annual working capital requirement of the dairy unit was estimated at Rs. 8,27,300. In the working capital requirement, the share of feed and fodder accounted for about 71 per cent of total working capital. Dairy farming is a labour intensive activity and the total annual expenditure on labour will be to the tune of Rs. 1,80,000 accounting for about 22 per cent of total working capital requirement.

Table 5: Economics of dairy farm of 10 cattle

Sr.no.	Particular	Amount (Rs)	Per cent
A	**Cost of production**		
1.	Fixed cost		
i	Interest on fixed investment @ 10 % /annum	89500	9.27
ii	Depreciation of shed @ 5% p.a.	25000	2.59
iii	Depreciation of utensils/implements @ 10 % p.a.	3500	0.36
	Total Fixed Cost (TFC)	118000	12.22
2.	**Variable Cost**		0
i	Total Working Capital	827300	85.64
ii	Interest on total working capital for half year @ 5% p.a.	20683	2.14
	Total Variable Cost	847983	87.78
3.	Total Cost (I+2)	965983	100
B.	**Production**		
i	Milk @2400 ltr./ cattle/year	24000	
ii	salable stock (heifers and adult animals replacement)	4	
iii	Farm Yard Manure (FYM) in q	228	
C.	**Returns**		
i	Income from milk @ Rs 45/L	1080000	89.58
ii	Sale of stock @ Rs 20,000/cattle	80000	6.64
iii	Value of FYM @ Rs 200/q	45600	3.78
iv	Gross returns (Rs)	1205600	
v	Net Returns (Rs/annum)	239617	
vi	Net Returns (Rs/month)	19968.08	
D	Benefit cost Ratio (BCR)	1.25	

Costs and Returns from Dairy Unit

The level of profits from the dairy unit can be estimated by deducting the costs of production from the gross returns. The economics of dairy unit of 10 Jersey crossbred is presented in Table 5. It can be observed from the table that the total annual cost of maintenance of dairy unit is estimated at Rs. 9,65,983. The fixed cost of the unit is associated to the fixed assets which comprised the interest on initial investment and depreciation of fixed assets. The total fixed cost of the unit is estimated at Rs. 1,18,000/annum accounting for about 12 per cent of the total cost. The major portion of the total cost is contributed by working capital. Based on the level of working capital used during the year, the variable cost of the unit was estimated at Rs. 8,47,983. The total annual gross income of the unit is estimated at Rs. 12,05,600 in which the share of income from milk, farm yard manure and sale of stock are 89.58, 6.64 and 3.78 per cent, respectively. The annual net returns from the dairy unit are Rs. 2,39,617 which indicates that the farmers will be able to generate a profit of worth Rs. 19,968/month. The benefit cost ratio of the unit is 1:1.25 which indicates that one rupee invested in dairy unit will generate an income of Rs. 1.25. Thus, it can be concluded that dairy farming is a profitable enterprise for the farming community of the state.

4. Protected Cultivation of Crops

As indicated earlier that the average size of the land holding in Himachal Pradesh is around one hectare. Due to the hilly topography the proportion of cultivated land is low and majority of the holdings are small and fragmented. In general, the crop production enterprises are undertaken in open conditions under natural conditions. In the existing scenario the crop production in open conditions is quite risky due to unforeseen circumstances like draught, heavy rains, hail storm, damages by stray and wild animals, birds, etc. In order to address these issues and reduce the production risk and uncertainties the protected crop production technologies are being promoted for the cultivation of high value crops. In this condition the production of crops is done in the protected structures commonly known as polyhouses. Since, construction of polyhouses is capital intensive and requires huge initial investment i.e. Rs 800-1000 per square metre depending upon the type of structure and quality of material used. The advantages of the protected cultivation over open cultivation are:

- The productivity of the crops is 3-4 times higher compared to the open cultivation.
- Protection of crops against natural calamities
- The quality of the produce is better than the open condition cultivations
- Facilitates proper management due to small operational area
- Less use of water due to micro irrigation systems
- Remunerative prices of produces on account of good quality and uniformity in the size and colour of fruits.
- Technology is suitable for small and marginal farmers to earn livelihood.
- The temperature inside the protected structures is 8-10° C higher compared to outside temperature, thus facilitates off-season production.

Keeping in view, the weak financial conditions of the farmers and promotion of protected cultivation of crops, state government is providing huge amount of subsidies under different schemes i.e. 80 per cent of the total cost of the structures. In general, the poly houses of the size 6, 40 and 100 square metres have been designed for the production of crop nurseries. Depending on the requirement of farming community the subsidies are available on the polyhouses of 6, 40, 100, 250, 500 and 1000 square meters sizes. Big polyhouses i.e. 250, 500 and 1000 square meters sizes are mainly used for crop production. The department of Agriculture, KVKs and SAUs are promoting the cultivation of capsicum, tomato and cucumber under protected cultivation.

Economics of Production of Vegetable Crops under Protected Conditions

As stated above the production of vegetable can successfully be undertaken in the polyhouses ofbig sizes. As far as utilization ofpolyhouse is concerned, it can be used for the cultivation of one crop of capsicum and two crops of tomato and cucumber each during a year. The economic analysis of the production of capsicum, tomato and cucumber in the polyhouse of 500square meters size presented in Table 6 indicated that the construction of polyhouse and associated facilities for irrigation etc. requires the initial investment of Rs. 5,42,000,with major investment of Rs. 4,50,000 on structure. The major portion of the initial investment is contributed by the government in the form of subsidy. The created infra-structure is for the cultivation of crops for a period of about 10 years, however, the sheet of polyhouse need replacement once or twice during the period if damaged by wind or animals etc. Thus, fixed cost of the structure has been estimated on the actual investment made by farmer. The total fixed cost on account of depreciation and interest on the investment is to the tune of Rs. 25,200 per annum. The variable cost of production of crops is estimated at Rs. 1,23,641, Rs. 1,25,947 and Rs. 1,47,725 in case of capsicum, tomato and cucumber, respectively. Among the various items of variable cost, the proportion of expenditure on human labour is highest (about 60%) followed by seed (13.54%) and inorganic fertilizers (7.11%) in cucumber. The total cost of production is highest in cucumber (Rs.1,72,928) followed by tomato (Rs. 1,51,147) and capsicum (Rs. 1,48,841).

Table 6: Economics of crop production under protected conditions (Rupees/annum)

S. No.	Particulars	Capsicum	Tomato	Cucumber
A.	**Initial Investment (Rs)**			
	Investment on the construction of polyhouse (500^2 size)	450000	450000	450000
	Irrigation infra-structure (Micro-Irrigation System	50000	50000	50000
	Water harvesting tank (9^3 metres	42000	42000	42000
	Total initial investment	542000	542000	542000
	Government subsidy	416000	416000	416000
	Farmer's share	126000	126000	126000
B	**Fixed Cost (Rs)**			
	Interest on farmer's initial investment @ 10% p.a	12600	12600	12600
	Depreciation of fixed assets @10% p.a	12600	12600	12600
	Total fixed cost (TFC)	25200	25200	25200
C	**Variable cost**			

S. No.	Particulars	Capsicum	Tomato	Cucumber
	Seed	2000	3000	20000
	FYM	2500	5000	5000
	Inorganic fertilizers	5625	6250	10500
	Plant protection	7500	6250	6250
	stacking	2500	1875	1875
	Soil treatment	5000	5000	5000
	human labour	89250	89250	89250
	Miscellaneous expenditures	6250	6250	6250
	Interest on total working capital @5% for 6 months	3016	3072	3603
	Total Variable cost	123641	125947	147728
D.	**Total Cost**	**148841**	**151147**	**172928**
E.	Production and Returns			
1.	Total Production (q)	60	85	120
2.	Average Price (Rs/q)	4000	2300	1900
3.	Gross Returns (Rs)	240000	195500	228000
4.	Net income (3-D) in Rupees	91159	44353	55072
F.	Benefit cost ratio	1.61	1.29	1.32

The data further reveal that during a period of one year a polyhouse of 500 squaremetres size produces 60 or 85 or 120 q of capsicum, tomato and cucumber, respectively. The average price received by the farmers was found to vary from Rs. 1900/q in cucumber to Rs. 4000/q in capsicum. Based on the gross returns and total cost of production, the net returns from capsicum are to the tune of Rs. 91,159 followed by cucumber (Rs. 55,072) and tomato (Rs. 44,353). The benefit cost ratio also indicates the similar pattern of income generation potential from these crops. The analysis indicates the cultivation of crop under the polyhouses a profitable activity especially for households who have small and marginal size of holdings and has problem of stray and wild animals.

Conclusion

The majority of the land holdings in the state are marginal and small and the average size of holdings is around one hectare. Thus, thelarge-scale production of the crop enterprises is not possible. The income of the farmers may be increased through the adoption of agri-based allied enterprises which are less land demanding. The enterprises discussed above (mushroom production, backyard poultry, small scale dairy and protected cultivation of crops) are the potential enterprises for doubling the farmers' income. The economic analysis of these enterprises revealed that round the year production of mushroom (three crops i.e. button, *dhingri,* milky mushroom of 200 bags unit each) has potential of generating a profit of Rs. 70000-80000/annum. The income from the cultivation vegetable crops under polyhouse condition generates a net income of Rs. 44000 (tomato) to Rs. 91000 (Capsicum).

Similarly, the profits from the small dairy unit (10 cattle) and backyard poultry unit of 25 *Himsamridhi* birds are to the tune of Rs. 2.39 lakh/annum and about Rs. 18000/annum, respectively. Therefore, atleast one activity amongst these can be promoted among the farmers of the state to increase the household income.

References

Annual Report (DAHDF), 2016-17, Department of Animal Husbandry, Dairying and Fisheries. Ministry of Agriculture, Govt. of India, New Delhi.

Baruah MS, Raghav CS and Kalita H. 2018. Effect of Technological Intervention on the Economics of Vanaraja Chicken Rearing in West Siang District of Arunachal Pradesh, India Journal of World's Poultry Research, 8(2): 44-49

Chand R, Saxena R and Rana S. 2015. Estimates and Analysis of Farm Income in India, 1983-84 to 2011-12. Economic and Political Weekly: 32

Chitra K, Venkatesh R, Dhanalakshmi K, Sharavanan PT, Sasikumar BC and Vijayakumari KK. 2018. Production and Economic Analysis of Oyster Mushroom (Pleurotusflorida). International Journal of Current Microbiology and Applied Sciences, 7(9): 379-383

Deka P, Sarma M, Nath PJ, Borgohain R, Mahanta JD, Deka B and M Phukon. 2014. Production performance of Vanaraja bird under traditional system of rearing in Assam. International Journal of Livestock Research. 4(2): 81-85

Evenson RE, Pray C and Rosegrant MW. 1999. Agricultural Research and Productivity Growth inIndia. Research Report No. 109. International Food Policy Research Institute, Washington, D.C.

GoI (Government of India). 2007. Agriculture Strategy for Eleventh Plan: Some Critical Issues. PlanningCommission, New Delhi.

Islam R, Nath P, Bharali A and Borah R. 2015. Analysis of benefit-cost (B:C) ratio of Vanaraja and Local chicken of Assam under backyard system of rearing. Journal of Research in Agriculture and Animal Science, 3 (7):7-10

Jaiswal P, Chandravanshi H, Netam A. 2018. Contribution of dairy farming in employment and household nutrition in India. International Journal of Avian and Wildlife Biology, 3(1)

Sahu J, Yadav A, Dinani OP, Kumari T, Dubey A and Gadpayle R. 2019. Role of Dairy Farming to Double the Farmers' Income by 2025. International Journal of Advances in Agricultural Science and Technology, 6 (2): 1-10

Sarial AK. 2016. Doubling farmers' income: A model for hilly and mountainous region. Himachal Journal of Agricultural Research, 42(2): 101-114

Singh DN, Sirohi R, Singh Y, Kumar A and Shukla PK. 2018. Role of Animal By-Products Utilization in Doubling the Farmer's Income.Indian Journal of Agriculture Business, 4(1): 23-34

Singh M, Islam R and Avasthe R. 2019. Socioeconomic impact of vanaraja backyard poultry farming in Sikkim Himalayas. International Journal of Livestock Research, 9(3): 243-248

Uddin MT, Islam MM, Salam S and Yasmin S. 2013. Economics of native poultry rearing in the coastal regions of Bangladesh.Bangladesh Journal of Animal Science, 42(1): 49-56

7

Attitude of Farmers Towards Kisan Credit Card Scheme in Etawah District of Uttar Pradesh

*Pankaj Kumar[1], Abhijeet[1], Rohit[1], Harshit Paliwal[2,] and Anjali Singh[3]

[1,3]*Department of Agricultural Extension, Chandra Shekhar Azad University of Agriculture & Technology Kanpur, Uttar Pradesh*
[2]*Department of Extension Education, Banaras Hindu University, Varanasi Uttar Pradesh*

Introduction

India pre-dominantly an agriculture-based country, requires constant endeavors to seewhether rural and agricultural credit facilities are enhanced with time. As a part of theseefforts, agricultural credit cards and cash credit facility were introduced in lending system. The Kisan Credit Card is a credit card programme launched by Indian banks in August 1998. The National Bank for Agriculture and Rural Development (NABARD) developed this model scheme to supply term loans for agricultural needs based on the suggestions of the R.V. Gupta Committee. Its goal is to provide financial assistance to farmers in order to address the agricultural sector's comprehensive credit needs. All commercial banks, Regional Rural Banks, and state cooperative banks are among the participants. Short-term credit limitations for crops are in place, as are term loans. Personal accident insurance covers KCC credit card members up to Rs. 50,000 for death and permanent disability and up to Rs. 25,000 for other risks. The premium is split evenly between the bank and the borrower. The validity period is five years, with the option of an additional three-year extension, Kisan Credit Card provides cash and term credit to farmers for a variety of linkedoperations such as pump sets, land expansion, planting, and drip irrigation. Soil Wellbeing Card programming has been institutionalized and online programming built for eight statesunderthee-Government Program to provide coordinated supplement administration suggestions based on soil test edit reaction approach. Ranchers receive data through avariety of channels, including Common Service Centres, Internet Kiosks, and SMS, as partofthe Nationale-Governance Plan for Agriculture. Farmers are spared from the exorbitant interest rates charged by traditional banks under the KCC plan. The interest rate on KCC loans ranges from 2% to 4%. This low interest rate aids farmers in repaying loans more quickly, taking into account the crop's harvesting season and the date the loan was provide

Methodology

The study was conducted in Etawah district of Utter Pradesh Two tahsils namely Etawah and Bharthana were selected randomly for study. Four villages Rahin, Kelamau, Birari, Ujhiyani, were selected from both tahsils. For this study 120 respondents were selected by simple random sampling procedure. A well-structured interview schedule designed for study was used for collecting the data from respondents through personal interview method. The data collected from the respondents were edited, tabulated and analysed using suitable statistical measures such as frequency, percentage, range, mean, standard deviation and coefficient of correlation were used.

Results and Discussion

The study was conducted purposively in Etawah district thus total 120 respondents were selected for the study. The study revealed that majority (60 %) of the respondents belonged to middle age group (i.e. 36 to 55 years), educated up to secondary education(30%), possessed small land holding (35.83%), had medium level of farming experience (62.50%), had low income (84.16%), had medium level of social participation (50.83%), had medium level extension contact (52.50 %), high level of achievement motivation (77.50%), had medium risk bearing ability (66.66%), had medium level of repayment behaviour (63.34 %), respondents had non-defaulter (70.83%), had Paddy , Bajra, Moong and Pigeon pea + Sunflower cropping pattern in Kharif season, Wheat, Mustard, Gram, and Potato cropping pattern in Rabi season, Groundnut, Moong, Bajra and Maize cropping pattern in Zaid season. Most of the respondents had moderately favourable attitude towards Kisan Credit Card Scheme in which Education, farming experience, cropping pattern, Annual income, Extension contact, Achievement motivation and Risk orientation were found significant. Majority of them were satisfied with the Kisan Credit Card Scheme and wanted to continue with this programme.

Table 1: Distribution of farmers according to age, education, land holding, farming experience, cropping pattern (kharif, rabi, zaid), Income, Social participation, Extension contact, Achievement motivation, Risk orientation, Repayment behaviour and Status defaulter.

Characteristics	Category	frequency	percentage
Age	Young (Up to 35)	34	28.34
	Middle (36 to 55)	72	60.00
	Old (56 and above)	14	11.66
Education	Illiterate	10	08.33
	Primary Education (1st to 5th standard)	21	17.50
	Middle school (6th to 8th standard)	30	25.00
	Secondary education (9th to 10th standard)	36	30.00
	Higher Secondary Education (11th to 12th standard)	12	10.00
	Graduate and Post graduate	11	09.16
Land holding	Marginal (Upto 1 ha)	25	20.84
	Small (1.01 to 2 ha.)	43	35.83
	Semi-medium (2.01 to 4 ha.)	34	28.34
	Medium (4.01 to 10 ha.)	16	13.33
	Large (10.01 ha. and above)	02	01.66
Farming experience	Less (up to 11)	24	20
	Moderate (12 to 33)	75	62.50
	More (34 and above)	21	17.50
Cropping pattern (Kharif)	Paddy	35	29.16
	Pigeonpea+Maize	5	04.16
	Moong	10	08.33
	Bajra	15	12.50
Rabi	Wheat	45	37.50
	Mustard	20	16.66
	Gram	15	12.50
Zaid	Groundnut	08	06.67
	Moong	20	16.66
	Bajra	10	08.33
Perennial	Guava	02	01.67
Other crops	Pea	07	05.83
	Potato	09	07.50
Income	Low (Up to Rs. 80,000)	16	13.34
	Medium (Rs. 80,001 to 2,00,00)	101	84.16
	High (Rs. 2,00,001 and above)	03	02.50
Social participation	Low(up to 5)	52	43.34
	Medium(6 to 10)	61	50.83
	High(11 and above)	07	05.83

Characteristics	Category	frequency	percentage
Extension contact	Low(up to 5)	52	43.34
	Medium(6 to 10)	63	52.50
	High(11 and above)	05	04.16
Achievement motivation	Low(up to 21)	10	08.34
	Medium(22 to 25)	17	14.16
	High(26 and above)	93	77.50
Risk orientation	Low(up to 21)	28	23.34
	Medium(22 to 26)	80	66.66
	High(27and above)	12	10.00
Repayment behaviour	Low(up to 2)	26	21.66
	Medium(3 to 4)	76	63.34
	High(5and above)	18	15.00
Status defaulter	Defaulter	35	29.17
	Non-defaulter	85	70.83

Attitude of Farmers Towards *Kisan Credit Card Scheme*

Attitude are frequency described in terms of personal consistency, as a "latent concept that is individual-dependent". Attitude strength is an important determinant of attitude behaviour relationship The attitude construct, as a person's degree of evaluate affect toward target behaviour. Attitude as a psychological tendency that is expressed by evaluating a particular entity with some degree of favour or disfavour. Individual might hold multiple attitudes about an object. The information regarding the attitude of farmers towards *Kisan Credit Card Scheme* was collected, tabulated and analyzed. The results are presented in

Table 2: Distribution of farmers according toover all attitude of farmers towards *Kisan Credit Card Scheme*.

Sr.No.	Category(Scores)	Respondents(n=120)	
		Frequency	**Percent**
1.	Less favorable(Upto101)	12	10.00
2.	Moderately favorable(102to121)	94	78.33
3.	Highly favorable(122andabove)	14	11.67
	Total	**120**	**100.00**

Table 3: Distribution of the respondents according to their attitude towardsKisan Credit Card Scheme.

S.No.	Statements	MPS	Rank
1.	This scheme is a boon for the farmers.	4.475	XX
2.	This scheme is beneficial for high agricultural production.	4.525	XXI
3.	I think this scheme motivates farmers to use improved technique.	3.8833	VII
4.	I don't like getting credit under this scheme because of the complicated process.	2.975	VI

S.No.	Statements	MPS	Rank
5.	I think this scheme gives the more stability to the farm business.	4.025	XIII
6.	This scheme prevents the family from financially deterioration.	3.95	XI
7.	The scheme provides credit facility to farmers in each season.	3.975	XII
8.	I think that all village farmers not included in the scheme.	2.10833	III
9.	State agricultural department provided me proper and sufficient information of this scheme.	4.13333	XVI
10.	I think, due to this scheme tendency of farmers to borrow credit from lenders reduced.	4.55	XXIII
11.	This scheme helps to save the farmers capital.	4.058333	XIV
12.	There is no need to apply for every crop to get loan under this scheme.	4.65833	XXV
13.	I think there is no loan in time under this scheme.	2.091667	II
14.	This scheme loan can be repaid after the completion of season.	4.425	XIX
15.	The interest rate for loans under this scheme is low and reasonable.	4.3833	XVIII
16.	I think this scheme has improved the financial literacy of the beneficiaries.	3.925	IX
17.	I think this scheme benefits for the selectors and the needy farmers are deprived from it.	1.891667	I
18.	Due to this scheme the farmers' attraction to businesses has reduced.	4.641667	XXIV
19.	This scheme creates more employments.	3.891667	VIII
20.	This is a favorable source of credit.	4.53333	XXII
21.	Debt restructuring is satisfactory.	4.066667	XV
22.	I think there is a favorable period for debt repayment.	3.93333	X
23.	Not get promotional credit given by this scheme.	2.2	IV
24.	The loan repayment period of this scheme i.e.12 months which is less.	2.583333	V
25.	This scheme has the flexibility to make withdrawals and to buy sustainably.	4.291667	XVII

MPS = Mean Percent Score

Table 3 : Shows the distribution respondents according to their attitude of farmers towards Kisan Credit Card Scheme, *I think this scheme benefits for the selectors and the needy farmers are deprived from it,* mean percent score 1.891667 rank I, *I think there is no loan in time under this scheme.* mean percent score 2.091667 rank II, *The scheme provides credit facility to farmers in each season,* mean percent score 2.10833 rank III, *Not get promotional credit given by this scheme,* mean percent score 2.2 rank IV, *The loan repayment period of this scheme i.e.12 months which is less,* mean percent 2.583333 rank V, *I don't like getting credit under* *this scheme because of the complicated process,* mean percent 2.975 rank VI, *I think this scheme motivates farmers to use improved technique,* mean percent 3.8833 rank VII, *This scheme creates more employments* , mean percent 3.8833 VIII, *I think this scheme has improved the financial literacy of the beneficiaries,* mean percent 3.925rank IX, *I think there is a favorable period for debt repayment,* mean percent 3.93333 rank X, *This scheme prevents the family from financially deterioration,* mean percent 3.95 rank XI, *The scheme provides credit facility to farmers in each season,* mean percent 3.975 rank XII, *I think this scheme gives the more stability to the farm business,* mean percent4.025 rank XIII, *This scheme helps to save the farmers capital,* , mean percent4.058333 rank XIV, *Debt restructuring is satisfactory* mean percent2.2 rank XV, *State agricultural department provided me proper and sufficient information of this scheme,* mean percent 4.13333 rank 4.13333 XVI, *This scheme has the flexibility to make withdrawals and to buy sustainably,* mean percent 4.291667 rank XVII, *The interest rate for loans under this scheme is low and reasonable,* mean percent 4.3833rankXVIII, *This scheme loan can be repaid after the completion of season,* mean percent rank 4.425 XIX, *This scheme is a boon for the farmers,*mean percent 4.475 rankXX,*This scheme is beneficial for high agricultural production,* mean percent 4.525 rankXXI, *This is a favorable source of credit,* mean percent 4.53333rankXXII, *I think, due to this scheme tendency of farmers to borrow credit from lenders reduced,* mean percent 4.55rankXXIII, *Due to this scheme the farmers' attraction to businesses,* mean percent,4.641667rankXXIV, *There is no need to apply for every crop to get loan under this scheme,* mean percent,4.641667rankXXV,

Relationship between independent variable with attitude of farmers towards Kisan Credit Card Scheme

The correlation coefficient (r) between attitude with various independent variables is presentedin Table 4.

Table 4: Distribution of the respondents according to theirRelationship between Socio-economic and Communication profile of the farmers with their Attitude.

Sr.No.	Independentvariables	Correlationcoefficient(r)
1.	Education	0.292**
2.	Farmingexperience	0.273**
3.	Croppingpattern	0.280**
4.	Annualincome	0.191*
5.	Extensioncontact	0.402**
6.	Achievementmotivation	0.378**
7.	Riskorientation	0.280**

**= Significant at1 percent levelofprobability(0.271)

*=Significantat5percentlevel of probability (0.179)

Table 4 : Distribution of the respondents according to their Relationship between Socio- economic and Communication profile of the farmers with their Attitude,

education and attitude, the correlation coefficient (r = 0.292*) revealed that the education and attitude of respondents had positive and significant. Therefore, it is concluded that higher the education highly favourable was the attitude level. They can read the relevant literature and grasp modern techniques of Kisan Credit Card Scheme. All this might have resulted in higher correlation between attitude level and education. This finding is consistent with the finding of Madhe (2019),***farming experience and attitude,*** the correlation coefficient (r = 0.273**) indicated that the relationship between size of farming experience and attitude level of farmers was positive and statistically significant. This clearly shows that increase in farming experience, also increases the level of attitude. Respondents with larger the experience could afford to take benefits of Kisan Credit Card Scheme due to which farming experience might be established positive and significant relationship with attitude. This finding is consistent with the finding of Natarajuet *al.(*2019*),* ***cropping pattern and attitude,*** the data in the Table 4, revealed that there was a positive and highly significant correlation (r = 0.280**) between cropping pattern and attitude of farmers towards Kisan Credit Card Scheme. It can be said that growing crops in different season encourages the farmers attitude towards Kisan Credit Card Scheme,this might be the reason that cropping pattern could express the positive and highly significant relationship with attitude. The same findings reported by Madhe (2019), ***annual income and attitude,*** the data in the Table 4, revealed that there was a positive and significant correlation (r = 0.191*) between annual income and attitude of farmers towards Kisan Credit Card Scheme, from this result it can be said that the annual income determines the economic status of the respondents. This clearly helps in developing the attitude of farmers. Annual income of the respondents therefore, could establish positive and significant relationship with attitude, the findings reported by Abu Razeim*et al.* (2017), ***extension contact and attitude,*** the data in the Table 4, revealed that there was a positive and highly significant correlation (r = 0.402**) between extension contact and attitude of farmers towards Kisan Credit Card Scheme. This result is consistent with Singh's (1999) findings, which showed that farmers' attitudes were favourably and significantly related to their involvement in extension wing operations. It is obvious that respondents with higher level of contact with extension agency are generally favourably predisposed to acquire more information, skills and other factors relating to enterprises consequently level of attitude. Thus, in this study positive and highly significant relationship must have been established between contact with extension agency and attitude. The findings of the study were in accordance with the Aaryaa Raaj and Jahanara (2017), ***achievement motivation and attitude;*** It is clear from the data depicted in the Table 4. that achievement motivation had positive and highly significant correlation (r = 0.378**) with attitude of respondents towards Kisan Credit Card Scheme. From the findings, this could be justified that any person challenging the crisis or toward incidents, their drive for excellence will be high and they want to reach high scales in the profession. As the risk orientation of the respondents was high they had high level of achievement motivation. The findings are in line with finding of Divakar (2013), ***Risk orientation and attitude,*** the data in the Table 4, concluded that there was a positive and highly significant correlation (r = 0.280**) between risk orientation and attitude of farmers towards Kisan Credit Card Scheme. This indicates that the higher the risk orientation also higher the level of attitude towards Kisan Credit Card Scheme. This finding may be due to those respondents who had high risk orientation are psychologically prepared to try new practices with a view to make progress in Kisan Credit Card Scheme. It is therefore, risk orientation was found to be positively and highly significantly related to attitude of farmers towards Kisan Credit Card Scheme. The findings of the study were similar to that of Korde (2017).

Conclusion

Attitude of farmers towards Kisan Credit Card Scheme, to examine the Relationship between socio-economic and communication profile of the farmers and their attitude towards Kisan Credit Card Scheme, it is concluded that the selected variables viz., education, farming experience, cropping pattern, annual income, extension contact, achievement motivation and risk orientation were in positive and significant relationship with attitude.

References

Aaryaa Raaj and Jahanara (2017). Profile of farmers and attitude towards soil health card scheme: a measure for the maintenance of soil health. Int. journal of res. Cul. society. Vol 1.

Abu Razeimet al.(2017) Attitude of farmers towards Pangas farming for their livelihood improvement. Asian Australas. J. Biosci. Biotechnol. 2 (1), 106-119.

Kumar, Daleep, Kalla, P.N. and Dangi, K.L. (2006). A study on attitude of farmer towards various activities of KVK. Raj. J. Ext. Edu.,14:106-108.

Korde V. M. (2017). Attitude of farmers towards organic farming M.Sc.(agri.) thesis, VNMKV Parbhani - 431 402 (M.S.), INDIA.

Mankar,D. M., P. P. Wankhade .and Y. B. Shambharkar. (2013). Impact of national horticultural mission on its beneficiaries. Int. J. Extn. Educ. 9 (1): 72-80.

Natraju M.S. (2019) Attitude of Farmers towards, agricultural Int. J. Curr. Microbiol. App. Sci. 8 (1): 2729-2734.

Sharma, R.K. and Bairathi, R. (1999). Attitude of farmers towards improved animal husbandary practices. PRASARIKA (Raj. J. Ext. Edu.)7:27-31.

Singh, G.(1999). Impact of Extension wing on the farmers. Rural India, 63 (3): 77-81.

Singh.P. Meenakshi choudhary and J.P. Lakhera(2014). Knowledge and Attitude Farmers towards improved wheat production technology Indian Res. J .Ext. Edu.14 (2): 110-115.

Sou and Mahendra (2022). Knowledge of farmers regarding isabgol (Plantago ovata) production in Jodhpur region of Rajasthan. Journal of Krishi Vigyan10 (2): 216-221.

8

Knowledge of Farm Women in Dairy Occupation

***R. U. Chaudhari**[1] **and U. M. Patel**[2]**

[1]Extension Education, CPCA, SDAU, S. K. Nagar, Gujarat
[2]Department of Veterinary Extension Education, Veterinary College KU, Anand

Abstract

Among the various animal husbandry practices the dairying has been playing a major role by providing source of revenue to rural people. Dairying is considered as a "wealth" of the Indian rural economy. It provides profitable service to a vast majority of rural family. The present study was performed in Gujarat. The analysis is concerned with Banaskantha region of Gujarat. The sample consisted of 150 farm women. Frequency and percentage were used for analysis of data. From banaskantha district, four talukas were selected randomly. From selected three taluka, four villages each and from one taluka three villages were selected constituting 15 villages and ten farm women were selected randomly from each village having dairy occupation. From this analysis it can be concluded that majority of farm women belonged to middle age group, primary school education, medium level of experience in dairy occupation, medium size of family, membership in one organization, marginal farmer, small herd size, medium level of source of information, medium level of milk yield. While, majority of farm women were found with medium level of knowledge in dairy occupation. Knowledge of farm women was found significant with age, whereas size of land holding and source of information were highly significant correlated with the knowledge of the farm women in dairy occupation. Experience in dairy occupation, social participation, herd size and milk yield were found non-significant. Whereas, education and size of family were found negative and non-significant.

Keywords: Knowledge, Farm Women, Relationship, Dairy Occupation, Profile.

Introduction

A woman is the nucleus of a family, particularly in rural India. For the rural women, the day starts early in the morning with the responsibilities of fetching water, fodder and fuel, cooking food and looking after the livestock and children. Promila *et al.* (2003) revealed that farm women spent 15:64 hrs/day and 16:58 hrs/day for productive work during the slack and peak period, respectively. The time saved by farm women from farm activities during the slack period was diverted towards care of children, family members and household works. Still, there is no recognition for their hard work; just because their work is not evaluated in terms of money.

The contribution of women in agriculture is estimated to be about 50 to 60 per cent. In many places, the entire livestock management is looked after by women only. They help in farm operation, take their animals on distant lands for grazing, washing, milking, fodder, breeding management, sale of milk, if necessary by taking it to the market and so on in addition, perform the functions related to house management. Even though, women are playing a very important role in innovation decision process, they are not allowed to try and see whether they are capable of managing farm and family affairs independently or with the help of family members.Adequate attention is also not being paid for the social and economic development of farm women and to encourage them to avail training facilities on farm and dairy management practices for supplementing their family income through development of occupational skills and proficiency.

Farm women's knowledge and skill in dairy occupation affect their efficiency in work in the development in dairy occupation. Therefore, felt necessary to determine "Knowledge of farm women in dairy occupation".

Objectives

1. To study profile of farm women.
2. To study the knowledge of farm women in dairy occupation.
3. To ascertain the relationship between profile of farm women and their knowledge in dairy occupation.

Methodology

The present study was conducted in Banaskantha district of Gujarat state during August 2019 to August 2021. From Banaskantha District, four talukas were selected randomly. From selected three taluka, four villages each and from one taluka, three villages were selected constituting 15 villages and ten farm women were selected randomly from each village having dairy occupation making a sample of 150 farm women. Interview schedule was prepared on the basis of objective on the subject. The data were collected through face-to-face interview. An ex-post-facto research design was applied.

Level of Knowledge

It can be defined as body of understood information possessed by an individual or by a culture. Knowledge regarding recommended dairy occupation was measured with the help of teacher made knowledge test based on the scale developed by Jha and Singh (1970) and was administrated with slight modifications to suit the present study.

The knowledge index was calculated with the help of following formula.

$$Knowledge\ index\ (Ki) = \frac{X_1 + X_2 + X_3 + \ldots\ldots\ldots + X_n}{n} \times 100$$

Where

X1 +X2 + X3 +.......... + Xn = Total number of correct answers i.e. total score

N = Total number of items in the test.

Results and Discussion

Based on information collected from farm women, the results obtained presented as per objectives are below.

1. Profile of farm women

Table 1: Distribution of farm women on the basis of their personal profile

(n = 150)

Sr. No.	Personal profile of farm women	Category	Frequency	Percentage
1	Age	Young age group (Up to 35 years)	41	27.33
		Middle age group (36 to 50 years)	92	61.34
		Old age group (Above 50 years)	17	11.33
		Total	**150**	**100.00**
2	Education	Illiterate (Do not read and write)	54	36.00
		Primary (Up to VIII standard)	69	46.00
		Secondary (IX to X standard)	19	12.67
		Higher secondary (XI to XII standard)	08	05.33
		College and above	00	00.00
		Total	**150**	**100.00**
3	Experience in dairy occupation	Less experience (Up to 10 years)	30	20.00
		Medium experience (Between 11 to 20 years)	105	70.00
		High experience (Above 20 years)	15	10.00
		Total	**150**	**100.00**

Sr. No.	Personal profile of farm women	Category	Frequency	Percentage
4	Size of family	Small family (Up to 4 members)	66	44.00
		Medium family (5 to 8 members)	79	52.67
		Large family (Above 8 members)	05	03.33
		Total	**150**	**100.00**
5	Social participation	No membership	05	03.33
		Membership in one organization	131	87.34
		Membership in more than one organization	14	09.33
		Holding position in organization	00	00.00
		Total	**150**	**100.00**
6	Size of land holding	Landless (No land)	03	02.00
		Marginal farmer (Up to 1.0 ha)	57	38.00
		Small farmer (Between 1.01 to 2.0 ha)	48	32.00
		Medium farmer (Between 2.01 to 4.0 ha)	32	21.33
		Big farmer (Above 4.0 ha)	10	06.67
		Total	**150**	**100.00**
7	Herd size	Small (Up to 5 animals)	94	62.67
		Medium (6 to 10 animals)	40	26.67
		Large (11 to 15 animals)	11	07.33
		Very large (More than 15 animals)	05	03.33
		Total	**150**	**100.00**
8	Sources of information	Low level of source of information	38	25.33
		Medium level of source of information	83	55.34
		High level of source of information	29	19.33
		Total	**150**	**100.00**
9	Milk yield	Low level of milk yield	19	12.67
		Medium level of milk yield	111	74.00
		High level of milk yield	20	13.33
		Total	**150**	**100.00**

From the above discussion, it can be concluded that majority of farm women belonged to middle age group (36 to 50 years), primary school (up to VIII standard), medium level of experience (between 11 to 20 years), medium size of family (5 to 8 members), membership in one organization, marginal farmer in size of land holding (up to 1.0 ha), small herd size (up to 5 animals), medium level of source of information, medium level of milk yield. The finding has been reported by Gamit *et al.* (2016), Hagone and Basunathe (2015), Islam *et al.* (2019), Kaur *et al.* (2019), Kimaro *et al.* (2013), Patel *et al.* (2017), Saba *et al.* (2020), The singuraja *et al.* (2017), Vekariya*et al.* (2016), Yadav and Revanna (2017).

2. Knowledge of farm women in dairy occupation

In order to determine the knowledge level of farm women in dairy occupation they were categorized into the following three groups.

Table 2: Distribution of farm women on the basis of their level of knowledge

(n = 150)

Sr.No.	Level of knowledge	Frequency	Percentage
1	Low level of knowledge	20	13.33
2	Medium level of knowledge	97	64.67
3	High level of knowledge	33	22.00
	Total	**150**	**100.00**
	$\bar{X}$= 46.99	S.D. = 08.68	

Farm women were found with medium level of knowledge in dairy occupation (64.67 per cent), whereas 22.00 and 13.33 per cent farm women were having with high and low level of knowledge in dairy occupation, respectively.

The probable reason could be that even though majority of farm women were primary educated and illiterate to in the study area, their ancestor had well know-how to rear their animal wisely/profitable, they learnt from them gradually. Also, frequent visit of officials of BANAS dairy and state veterinary department gave timely understandable and updated information required to farm women for profitable dairy occupation which definitely influences their knowledge level. This finding has been supported by Patil *et al.* (2009) and Parmar and Sharma (2014).

3. Relationship between profile of farm women and their knowledge in dairy occupation

In order to find out the relationship between profile (Age, Education, Experience in dairy occupation, Size of family, Social participation, Size of land holding, Herd size, Sources of information, Milk yield) of farm women and their knowledge in dairy occupation, coefficient of correlation was worked out. The data in this regard are presented in Table 3.

Table 3: Relationship between profile of farm women and their knowledge

(n = 150)

Sr. No.	Independent variables	Correlation coefficients (r values)
1	Age	0.190^{*}
2	Education	-0.091^{NS}
3	Experience in dairy occupation	0.101^{NS}
4	Size of family	-0.094^{NS}
5	Social participation	0.147^{NS}
6	Size of land holding	0.290^{**}
7	Herd size	0.120^{NS}
8	Source of information	0.449^{**}
9	Milk yield	0.073^{NS}

* Significant at 0.05 level of probability
** Significant at 0.01 level of probability and
NS: Non Significant.

Out of nine independent variables, the independent variable like age was significantly correlated with the knowledge whereas size of land holding and source of information were highly significant correlated with the knowledge of the farm women in dairy occupation. Experience in dairy occupation, social participation, herd size and milk yield were found non-significant. Whereas education and size of family were found negative and non-significant.

Conclusion

From this analysis it can be concluded that majority of farm women belonged to middle age group followed by primary school education, medium level of experience in dairy occupation, medium size of family, membership in one organization, marginal farmer, small herd size, medium level of source of information, medium level of milk yield. While, majority of farm women were found with medium level of knowledge in dairy occupation. Knowledge of farm women was found significant with age, whereas size of land holding and source of information were highly significant correlated with the knowledge of the farm women in dairy occupation. Experience in dairy occupation, social participation, herd size and milk yield were found non-significant. Whereas, education and size of family were found negative and non-significant.

References

Gamit, M. P.; Rani D. V.; Bhabhor, I. N.; Rathod, A. D. and Vyas, H. U. (2016). Personal, socio-economic and psychological characteristics of dairy farmers in Surat District of South Gujarat. Advances in Life Sciences,5(1): 45-49.

Hagone, V. and Basunathe, V. K. (2015). Decision making pattern of tribal women in dairy enterprise in Melghat region of Amravati district. Indian Research Journal of Extension Education,15(2): 91-94.

Islam Md. S.; S. S.; Islam, Md. M. and Billah, Md. M. (2019). Exploring women participation in small-scale dairy farming: a case of PaikgachhaUpazila, Khulna, Bangladesh. Asian Journal of Research in Animal and Veterinary Sciences, 3(4): 1-13.

Jha, P. N. and Singh, K. N. (1970). A test to measure farmer's knowledge about high yielding varieties programme. Inter Discipline,7(1): 65-68.

Kaur, S.; Singh, J.; Verma, H. K.; Dash, S. K. and Kansal, S. K. (2019). Participation Appraisal of Women Farmers in Dairy Husbandry Practices in Indo-Pak Border Area of Punjab (India). International Journal of Current Microbiology and Applied Sciences, 8(5): 2298-2305.

Kimaro, E. G.; Lyimo-Macha, J. G. and Jeckoniah, J. N. (2013). Gender roles in small holder dairy farming : pertinent issues on access and control over dairy farming resources in Arumeru. Livestock Research for Rural Development,25(5): 1-8.

Parmar, V. S. and Sharma O. P. (2014). Socio-personal characteristics of farm women in dairy occupation. Gujarat Journal of Extension Education,25 (1): 47-48.

Patel, S. J.; Kumar, R.; Patel, A. S.; Patel, N. R. and Parmar, V. N. (2017). Involvement of farm women in decision making regarding dairy farming in Junagadh District of Gujarat State. Indian Journal of Hill Farming, 30(I): 45-58.

Patil, A. P.; Gawande, S. H.; Nande, M. P. and Gobade, M. R. (2009). Assessment of knowledge level of dairy farmers in Nagpur district and the co-relation between socio-economic variables with their training needs. Veterinary World,2(5): 199-201.

Promila, K.; Yadav, D. and Sharma, N. (2003). Time use pattern of hill farm women, a study in Himachal Pradesh. Himachal Journal of Agricultural Research, 29(1-2): 84-88.

Saba; Akhtar, S.; Khalid, W. and Khan, S. (2020). Role of women in livestock management in the rural area of district Swabi, Khyber Pakhtunkhwa, Pakistan.Sarhad Journal of Agriculture,36(1): 291-302.

Thesinguraja, S.; Mathialagan, P.; Thilakar, P.; Devendran, P. and Palanichamy, V. (2017). Socio-economic profile of pulikulam cattle rearers in Madurai and Sivagangai Districts of Tamil Nadu, India. International Journal of Current Microbiology and Applied Sciences, 6(12): 424-429.

Vekariya, S. J.; Kumar, R., Savsani, H. H.; Kotadiya, C. R.; Chaudhari, D. M. and Chatrabhuji, B. B. (2016). Socio-Economic Profile of Maldhari Dairy Farmers of South Saurashtra Region. Current Agriculture Research Journal,4(2): 186-190.

Yadav, K. D. K. and Revanna, M. L. (2017). A study on socio-economic status of the farm women of Tumakuru District of Karnataka State, India. International Journal of Pure Applied for Biosciences, 5(4): 309-314.

9

Decision Making Power of Rural Women in Farm and Dairy Activities in Bhadrak District of Odisha

***Sweta Sahoo*[1*]*, Ashish Anand*[2]*, Smaranika Mohanty*[3]**
***Soumya Sakti Das*[4]**

[1]*Department of Agricultural Extension, IAS, SoA, Bhubaneswar, Odisha*
[2]*Department of Extension Education, CoA,OUAT, Bhubaneswar, Odisha*
[3]*Department of Agricultural Extension, IAS, SoA, Bhubaneswar, Odisha*
[4]*Department of Agricultural Extension, IASD, SoA, Bhubaneswar, Odisha*

Abstract

Women make up 40 per cent of the agricultural labour force in most parts of the world, while in developing countries they make up 67 per cent of the agricultural workforce. Therefore, strengthening and expanding women's roles and opportunities in agriculture promotes sustainable economic development. In conservation issues and sustainability, gender differentials are most obvious in rural areas. Regarding this,present study was conducted in Bhadrak district of Odisha to assess the decision making power of women in farm and dairy activities. For this, primary data has been collected from 120 respondents through personal interview.It was found that 'Weed management' aspect of decision making secured the first position and 'Adoption of new dairy practises' held last position among sixteen identified criterias regarding different aspects in decision making. With respect to decision making power,it was observed that majority of the respondents i.e. 64.16 per cent had medium level of decision making power followed by 19.17 per cent having low level and 16.67 per cent having high level of decision making power.This is because we have been living in a patriarchal society where all rights on decision making are always on men. These results can be utilized as feedback information to plan and execute the programmes for women empowerment in study area, particularly, where the contribution of women found to much higher.

Keywords: Decision making, Empowerment, Feedback, Gender differentials, Sustainable

Introduction

"If you teach a man to farm, his family will eat.
If you teach a woman to farm, the community will eat."

Women make essential contributions to the agricultural and rural economies in all developing countries. Their roles vary considerably between and within regions and are changing rapidly in many parts of the world, where economic and social forces are transforming the agricultural sector. Agriculture needs manpower, if the manpower split into gender wise, amazing fact is that the women contribution is greater or equal to men (Chauhan, 2011).That is reason that right from the day the country become independent, it has been trying hard for rural development.

According to the Penguin Atlas of Women in the World, women make up 40 per cent of the agricultural labour force in most parts of the world, while in developing countries they make up 67 per cent of the agricultural workforce (Singh, 2008). Therefore, strengthening and expanding women's roles and opportunities in agriculture promotes sustainable economic development. Gender inequalities are a measurable and significant limit on agricultural productivity and efficiency, undermining rural economic development. India has paved the way to development by increasing its trade and growth in Agriculture sector. To review the agriculture scenario of our country, it depicted that there is need to pay sincere attention towards the human resource development (men and women together) because modernizing agriculture needs skill and efficient work forced. Farm business has become a family enterprise in India, where both men and women take part evenly. It is well known fact that the success of rural development process largely depends on the participation of people at large, irrespective of sex. The problems of involving women's participation in the development process are now catching the attention of planners and policy maker because of increasing imbalances generated out of development process. Now days, rural women play a vital role in domestic and socio-economic life of the society and therefore, national development is not possible without developing this important and substantial section of our society (Okafor, 2008).

Women comprise half of the work force in agriculture especially in developing countries. In India according to 2011 census, female population is 51.27 per cent of total population, out of which 61.72 per cent come from rural areas. Although, the legal and constitutional provisions for equal status, they have been traditionally put in a weak position in Indian society and have a subordinate role to play. In spite of this fact, women share most of the family responsibilities and perform wide range of duties on farm and home.

On the basis of fact and data based research, it can be stated that increasing number of rural women in India are not simply housewives but are in fact farmers. We have today plenty of empirical evidence to show that women do participate practically in all regions, crop zones and at different stages of agricultural production. Despite various handicaps like insufficient access to education and exposure to public life, they have shown sensitive response to changes and innovations. Of course, there are regional variations in women participation caused by factors including their heavy responsibilities in the production process, traditional believes and practices, caste restrictions, introduction of labour displacing technology, changes in production relation etc. These are not static and have to be seen within a dynamic process of changes (Madhushree, 2014).

Decision-making basically is a thinking process, which involves a process of discrimination from alternatives, possibilities involving many complex elements. In practices, the farming community does not take decision unilaterally or in isolation about several aspects of farming and home activities. But they have to end up with decision collectively at appropriate level in a rational manner. It has been pointed out by earlier authors that the decisions of farmers are influenced by the opinion of important persons with whom they interact. The adoption of an innovation is dependent on the decision making process (Akinkunmi, 2012).

Empirical evidences showed that women in rural area are more adversely affected by poverty than men. The incidence of poverty among rural women is on the rise in most of the developing countries and India is one of them. The issues of gender bias and equity point of the double burden women have to bear that on being poor and being women. Projects aiming to reduce poverty view the poor rural women as the recipient of benefits of development, instead of active participant and still poor rural women have the least access to basic needs such as food, health and education (Javed et al., 2016).

In nutshell, women have access neither to agricultural information and services nor to production assets. They have also no control over their earnings. This does affect their motivation to engage in different types of income generation activities. It is in this context that the involvements of women in production oriented training and extension programmers' not only as beneficiaries but also as change agents assumes great significance without education and training. They cannot have access to credit; technology and skills they need to better their lot and increase their efficiency as productive worker.

Non availability of time is an important reason for non-participation of women in various agricultural and home making activities. The present study aims at exploring the extent of participation of rural women in agricultural and home making activities. This will serve as feedback for strengthening the existing development programmers and planning and making then more meaningful.

Aim and Advantages of the Study

The present study is an attempt to find out the factors influencing participation and decision making of rural women in agriculture and constraints faced by them.The findings of the investigation of present micro level study will provide an overall picture about participation of rural women in agriculture and decision-making pattern of the study area in terms of their socio psychological personal and socio–economic attributes, and constraints perceived by them in decision-making.

Materials and Method

The present study was conducted in Bhadrak district of the Odisha state.Women prominently and actively participate in various farming activities and contribute in their household economy.There are 7 blocks functioning in the District. Bhadrak rural block was selected by purposive sampling for the research due to more number of female farmers in the Block as per pilot survey. There are 169 villages in Bhadrak Rural Block. Out of these, 10 villages were selected by purposive sampling for the present study based on number of women participating in farm and dairy activities.

From the 10 villages, 120 respondents were selected randomly. A structured interview schedule was used for data collection. Descriptive research design was used for this study. The statistical tools such as frequency, percentage,mean and standard deviation were used to interpret the data and for drawing logical conclusion.

Results and Discussion

Table 1: Distribution of respondents based on participation in decision making

Sl. no.	Activity	Always (F)	P	Sometimes (F)	P	Never (F)	P	Mean Score
1	Selection of crop	35	29.17	64	53.33	21	17.50	2.12(V)
2	Selection of crop variety	12	10.00	52	43.33	56	46.66	1.63(IX)
3	Adoption of new farm practices	8	6.67	38	31.66	74	61.66	1.45(XII)
4	Adoption of new dairy practices	6	5.00	23	19.16	91	75.83	1.29(XVI)
5	Where to buy inputs	11	9.17	43	35.83	66	55.00	1.54(XI)
6	Credit management	34	28.34	51	42.50	35	29.16	1.99(VIII)
7	Purchasing farm tools	4	3.34	31	25.83	85	70.83	1.33(XV)
8	Selection of fodder	53	44.17	59	49.16	8	6.66	2.38(IV)
9	Irrigation	13	10.84	48	40.00	59	49.16	1.62(X)
10	Weed management	81	67.50	32	26.66	7	5.83	2.62(I)
11	Fertiliser, insecticide management	9	7.50	29	24.16	82	68.33	1.39(XIII)
12	Harvesting	31	25.84	68	25.83	21	17.50	2.08(VI)

Sl. no.	Activity	Always (F)	P	Sometimes (F)	P	Never (F)	P	Mean Score
13	Where to sell produce	23	19.17	79	65.83	18	15.00	2.04(VII)
14	Where to store produce	83	69.17	26	21.66	11	9.16	2.60(II)
15	Which dairy product to prepare	79	65.84	23	19.16	18	15.00	2.51(III)
16	Insemination	8	6.67	29	24.17	83	69.16	1.38(XIV)

From the table 1,it can be seen that the decision of weed management held the most prominent position among sixteen farm activities quantified here followed by storing produce,dairy product to be prepared,fodderselection,selection of crop,harvesting,place of selling of produce,creditmanagement,selection of crop variety,irrigation,input buying place,adoption of new farm practices,fertilizer and insecticide management,insemination,purchasing farm tools and last among them was adoption of new dairy practices.

Table 2: Distribution of respondents based on level of decision making

Sl. No.	Category	Frequency	Percentage
1	Low (<27.54)	23	19.17
2	Medium (27.54-32.38)	77	64.16
3	High (>32.38)	20	16.67
	Total	120	100.00

The results portrayed in Table 2 show that majority of farmers belonged to the category having medium level of decision making i.e. 64.16 per cent followed by low level(19.17%) and high level(16.67%).

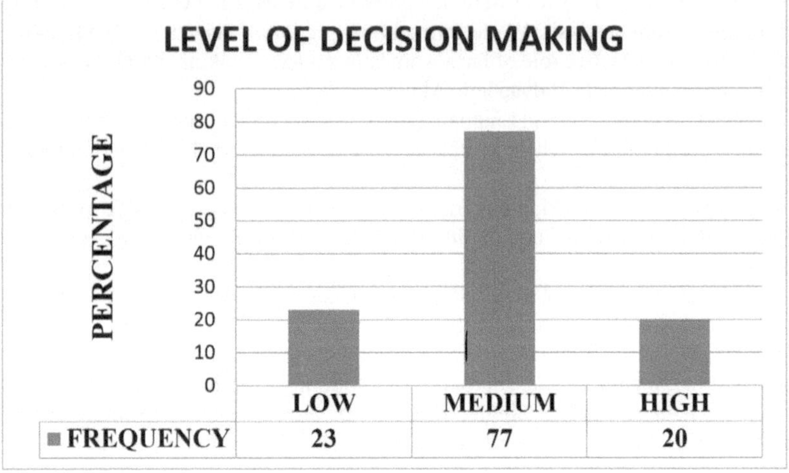

Fig. 1: Level of decision making of respondents

Conclusion

Present study focused on assessing the decision making power of women farmers. It was found that decision related to weed management held the first position among sampled women farmers. It was also observed from the study that majority of respondents i.e. 64.16 per cent had medium level of decision making.

Future Scope of the Study

The results of present study can be utilized as feedback information to plan and execute the programmes for women empowerment in study area, particularly, where the contribution of women found to much higher. In short, this study may be of a great value not only to extension workers but also to agricultural scientists, administrators, planners and policy makers towards attainment of national objectives of women empowerment, so that women will receive their due in the development process.

References

Abiola M.(2020) Extent of Involvement of Women in Dairy Farming Activities in Oyo State, Nigeria.Adisa,Agricultural Science,2(1).

Akinkunmi J. (2012). Assessing participation of women in poultry production as a sustainable livelihood choice in Oyo state, Nigeria. International Journal of plant, animal and environmental sciences,2(2):74-82.

Bhagirathi. (2016). Role of women in agricultural. Yojana, 44:35-38.

Chayal, K. and B.L. Dhaka. (2010). Analysis of role performance of women in farm activities. Indian Research Journal of Extension Education,10(2):109-112.

Chauhan, M. Nikulsinh. (2011). Role performance of tribal farm women in agriculture and animal husbandry in Gujrat. Karnataka journal of Agriculture sciences,24(5): 672-674.

Damisa, M.A and M. Yohanna. (2007). Role of rural women in farm management decision making process: ordered probit analysis. World Journal of Agricultural Sciences,3(4):543-546.

Ekale Jayashri. (2016). A perspective role of farm women in decision-making. Ph.D. Thesis, Dr. Punjabrao Deshmukh Krishi Vidyapeeth, Akola.

Javed Asif., Saima Sadaf and Muhammad Luqman. (2016). Rural Women's participation in crop and livestock production activities in Faisalabad-Pakistan. Journal of Agriculture and social sciences. 2(3):150-154.

Madhushree, A. (2014). Decision making pattern and achievement motivation of farm women under dry land condition in Tumkur district. M.Sc. (agri.) Thesis, Univ. Agri. Sci., Bengaluru.

Mishra, A., A. Mishra, and Dubey, A.K. (2009). Participation of rural women in decision making. Indian Research Journal of Extension Education. 9(3):23-25.

Okafor, C.N. (2008). Women participation in agricultural decision making in Aguto Local Government area, Anambra State. Journal of Agricultural Extension. 12(2): 108-113.

Singh, A.K., Jitendra Chauhan and S.C. Singh. (2008). Involvement behaviour of farm women in farm enterprises. Indian Research Journal of Extension Education. 8 (2 and3): 19-21.

10

Demographic Profile of Cauliflower Growers in Hardoi District of Uttar Pradesh

Aman Verma[1], Adesh Kumar Verma[2] , R.K. Doharey[3] N.R. Meena[4], Arvind Kumar[5], Anurag Dixit[6], Shyam[7]

[1,3,4,5,6,7]Department of Extension Education, Acharya Narendra Deva University of Agriculture & Technology, Kumarganj, Ayodhya, Uttar Pradesh
[2]Department of Agricultural Extension, Kulbhaskar Ashram Post Graduate College, Prayagraj, Uttar Pradesh

Abstract

The present paper attempts to examine the demographic profile of cauliflower growers in Hardoi district of Uttar Pradesh of India. The study was conducted in 10 villages with random selection of 120 respondents. The findings showed (43.33%) of respondents age between 35-50 years, earning Rs.80000-120000 (32.50%) annually, had marginal landholding (38.33%) and (74.17%) of the cauliflower growers are literate. With respect to farming, 75% respondents were reported agriculture and allied activities as their main occupation, the maximum no of respondent (44.17%) respondent having pakka house, (48.20%) were medium family size (5-7members) and (68.45%) were nuclear family type.

Introduction

Agriculture is the back bone of Indian economy. About 18.8 percent (2021-22) of the national income originates from the agriculture sector. Approximately 60 percent of India's population is directly or indirectly dependent on agriculture for livelihood. A large number of important industries like jute, textiles, edible oils, tobacco, sugar etc. receive the raw materials produced by agriculture sector. India occupies second position in the production of fruits and vegetables in the world after China (National Horticulture Database 2019-2020) It is grown almost in all the states of the country.

Cauliflower (*Brassica oleracea L. var. botrytis*) like all other Brassicaceae family crops has descended through introgression, mutation, human selection and adaptation from a single wild predecessor *Brassica oleracea* L. var. *sylvestris* which is still found growing in western and southern Europe and North Africa. The name cauliflower consists of two Latin words namely 'caulis' means cabbage and 'floris' means flower. It is grown throughout the country for its highly suppressed 'prefloral fleshy apical meristem' called "Curd" which is used as vegetable, soup and for pickling (Choudhury, 2006). Cauliflower is the pride of cole crops because of its good taste, flavour and nutritional value. It contains good amount of vitamins like vitamin A, vitamin C and fair amount of proteins and fibers. The cauliflower is also a good source of minerals like Ca, Mg, P, Fe, Na, and S.

It was rightly called as "Aristocrat of Cole crops". The crop is known to be indigenous of Mediterranean region and was introduced in India during 1822 from England (Hazra et al., 2011) during British time.

The popularity of cauliflower has been constantly increasing and world over it being cultivated over an area of 473 thousand hectares with annual production of 9283 thousand tonnes and productivity of 19.17 tonnes/hac (Horticulture Statistics Division, DAC&FW 2020-2021). In India, it is a major vegetable crop grown mainly in states like Bihar, Uttar Pradesh, Orissa, West Bengal, Assam, Haryana and Maharashtra etc.

Utter Pradesh has 9th position in thc production of cauliflower after West Bengal, Madhya Pradesh, Bihar, Gujrat, Haryana, Orrisa, Chattisgarh, Punjab, states. The area under cauliflower cultivation in Utter Pradesh is 20000 ha with production of 436.77 MT in (National Horticulture Board 2017-18). Vegetables play an important role in the maintenance of human health. These make diet nutritive and balanced. A balanced diet requires a proper quota of fresh vegetables. About 300 gm of vegetables are needed (90 gm root vegetables, 120gm green vegetables and 90gm other vegetables). But about 220.8 gm vegetables per day per capita are available (Indian Agriculture, 2001).

Cauliflower is highly nutritious vegetable. It is good source of minerals and vitamins. It is rich source of protein and carbohydrates, it provides 25 Kcal of energy. Cauliflower is low in fat, high in dietary fiber, folate, water and vitamin C, possessing a very high nutritional density.

Methodology: The present study was undertaken in Hardoi district of Uttar Pradesh. From them selected five block were selected purposely on the basis of highest area and production. Ten villages were selected from the selected block by using simple random sampling with proportional allocation to the size of selected village. A sample of 120 respondent farmers growing cauliflower at least in 1 *bigha* were selected from each selected village by using simple random sampling with proportion to the size of selected village.The pre-tested interview schedule was used for collection of data and the data was analysed by using appropriate statistical methods such aspercentage (%), mean and standard deviation.

Statistical Methods Used

The 'percentage' and 'average' SD and correlation coefficients 'r' were used for making simple interpretation.

1 Percentage (%)

The frequency of a particular cell was divided by the total number of respondents or (MPS) in that particular category and multiplied by 100 for calculating the percentage.

2 Average ($\bar{X}$)

The average ($\bar{X}$) was calculated by adding the total scores obtained by the respondents and divided itbythetotal number of respondents using the following formula:

$$(\bar{X}) = \frac{\sum X}{N}$$

Where,

$\bar{X}$ = Average or mean

$\sum X$ = Total number of scores obtained by respondents

N = Total number of respondents

3 Standard deviation (σ)

S.D. is the square root of mean of the squares of all deviations, the directions being measured from the arithmetic mean of the distribution. It is commonly developed by symbol (σ).

$$\text{S.D. } (\sigma) = \overline{X}$$

Where,

σ = Standard deviation

d = Deviation of variables mean

M = Total number of items

Results and Discussion: Socio-economic characteristics This section is devoted to the description of personal, socioeconomic and psychological characteristics of the respondents. It is in the fulfillment of the first objective set-up for the study.

1. **Age :** Age is a factor that can determines the level of dynamics and experience of an individual which affect an individual's ability to make decision and participate in the progress in his community.

Table 1: Age wise distribution of the respondents

Age (in years)	Frequency	Percentage
Young (<30)	31	25.83
Middle (30-50)	52	43.33
Old (>50)	37	30.84
Total	120	100.00

The table 1 revealed that (43.33%) respondents were is middle age followed by (30.84%) old age and (25.83%) per cent young age. It inferred that majority of the respondents are 30-50 age group.

2. **Caste:** Caste is a social status in which society placed an individual on the bases of his socio-economic characteristics.

Table 2: Distribution of the respondents on the basis of their caste

Categories	Frequency	Percentage
General	19	15.83
OBC	65	54.17
SC	34	28.33
ST	2	1.67
Total	120	100

It was found that (54.17%) per cent of vegetable farmers belonged to OBC caste category followed by SC (28.33%) and General (15.83%). Only 2 respondent was found from ST category.

3. **Educational attainment:** It is well known that education plays a vital role in human society it is recognized as one of the key instruments through which a society socialize its members and effects the desirable changes in the socio-economic life of its people.

Table 3: Distribution of the respondents on the basis of their education

Educational attainment	Frequency	Percentage
Illeterate	31	25.83
Primary school	19	15.83
Middle school	17	14.17

Educational attainment	Frequency	Percentage
High school	22	18.33
Intermediate	17	14.16
Graduate	12	10
Post graduate	2	1.67
Total	120	100

The table 3 indicated that most (18.33%) of respondents attended high school (25.83%) illiterate, (14.17%) middle school, (15.83%) primary school, (10%) graduate and (1.67%) post graduate respectively. It inferred that considerable percentages of the respondents had education up to high school and others were illiterates. Devarde (1981) reported that majority of the mango growers (77.92%) had education varying from primary to higher secondary level.

4. **Family Size**

Table 4: Distribution of the respondents on the basis of their family size

Size of family	Frequency	Percentage
Small (<4)	33	27.5
Medium (5-8)	67	55.83
Large (>8)	20	16.67
Total	120	100

It is evident from the table 4 that 55.83 percent respondents family were observed such who had (5-8) member followed by 27.5 percent family having (up to 4) members and 16.67 percent respondent family were found 8 and above member in their families. The range between minimum and maximum number of family members was recorded from 4 to 14.

Thus, it is concluded that the majority of the respondents were found in medium family size category.

5. **Family Type:** Individual behavior is often affected by types of family. Person who comes from nuclear family may behave differently with one who comes from joint family.

Table 5: Distribution of the respondents on the basis of their family type

Family type	Frequency	Percentage
Nuclear	71	59.17
Joint	49	40.83
Total	120	100

The distribution of respondent according to the family type shows that majority of the respondents (59.17%) having nuclear family and (40.83%) having joint family. The finding is in line with finding of Ninga Reddy (2005) reported that

62.67% of Beneficiaries belonged nuclear family. Whereas, 37.33% of them belonged to joint family. It can summarized that majority of the respondents belong to nuclear family.

6. Housing Pattern

Table 6: Distribution of the respondents on the basis of their housing pattern:

Housing pattern	Frequency	Percentage
Mix	53	44.17
Kutcha	24	20
Hut	2	1.67
Pakka	41	34.16
Total	120	100

It is apparent from the data shown in the Table 6, pertaining to type of house possession, that mixed type of habitation was observed to the extent of 44.17% followed by 34.16% pukka category and 20.00% kutcha and only 1.67% hut. So, it can be concluded that vegetable farmers were having better quality house. This may be due to good to socio-economic condition of the vegetable farmers, which is the result of good vegetable production. It also indicates the status of vegetable farmers in the society of the study area.

7. Land Holding: Land holding is an important indicator of socio-economic status of a family, size of land holding is closely associated with level of income and standard of living of the household.

Table 7: Distribution of the respondents on the basis of their land holding:

Land holding	Frequency	Percentage
Marginal (<1hac.)	58	48.33
Small (1-2hac)	35	29.16
Sub medium (2-4 hac)	13	10.83
Medium (4-10hac)	10	8.33
Large (>10hac)	4	3.33
Total	120	100

Study depicts that 48.33 per cent of farmers were having less than 1 ha of land, thus belonged to marginal farmers category. The farmers who belonged to small and medium categories were 29.16 per cent and 10.83 per cent, respectively. Data also shows that only 3.33 per cent of vegetable farmers were having large land holding. Thus, it may be concluded that majority of the vegetable farmers were marginal farmers having less than 1 hectare of agricultural land. This is due to the fact that in midden Uttar Pradesh, per capita agricultural land is comparatively less. The other reason may be the fragmentation of the holdings due to nuclear family system.

8. Occupation

Table 8: Distribution of the respondents on the basis of their occupation

Occupation	**Frequency**	**Percentage**
Farming	47	39.17
Farming+Animal husbandry	43	35.83
Government service	3	2.5
Others	27	22.5
Total	120	100

The distribution of respondents showed in the table 8 the respondents were having farming 39.17% as their main occupation followed by 35.83% farming and animal husbandry, 22.5% others occupation, 2.5% government services. It inferred that majority of the respondents depend on farming for day to day earning. It can be concluded that most of respondents were farmers and having farming as their main occupation.

9. Annual Income: Income plays a significant role in decision making, adoption of new ideas and participation in economic activities of an individual.

Table 9: Distribution of the respondents on the basis of annual income

Income status (Rs)	**Frequency**	**Percentage**
Up to 40000	24	20.00
40000-80000	33	27.50
80000-120000	39	32.50
More than 120000	24	20.00
Total	120	100

The table 9 showed that 32.5% of the respondents were earning Rs. 80000-120000, followed by 27.50%, 20% and 20% earning Rs. 40000-80000, more than Rs. 120000 and up to Rs. 40000 respectively. It inferred that majority of the respondents are living below or on the poverty line.

10. Decision Making Ability

Table 10: Distribution of respondents according to their decision making ability

Category	**Frequency**	**Percentage**
Individual decision	33	27.5
Joint decision with family members	58	48.33
Joint decision with other than family members	29	24.17
Total	120	100

It could be observed from Table 10, that near about half of the respondents (48.33%) made joint decisions by having consultation with their family members followed by (27.50%) of respondents who made individual decisions. A smaller proportion (24.17%) made joint decisions with members of other than their family members.

Conclusion

Based on the major findings of the study, it was concluded that the majority of the respondents were having medium socio-economic background and medium level in information management behaviour based on their annual income. Information management behavior of individual respondents was positively influenced by individual educational level, annual income, occupation, and market orientation. Also, the study concluded that most of the respondents were ready to accept new technological innovation to boost up their production due to their medium level of adoption, the major constraints faced by the respondents were unawareness about skill to adopt recommended technology, untimely availability of critical inputs and irregular field visits by extension personnel for providing information timely and their major constraints could be solved by providing them appropriate skills about latest recommended technologies of cauliflower production.

References

Choudhary A., M.K., Patel, P.C., Sharma, P.K. and Patel, J.B. (2010)..a study on socio economic profile of respondents. International Journal of Agriculture Sciences, 8(25):1521 -1523.

Dan alan, Lokesh, G.B. Amrutha, Joshi, T. Reddy b.S.And Hulagur, BB (2020). Socio economic Performance of Pradhan Mantri Fasal Bima Yojana(PMFBY) in Hyderabad-Karnataka (H-K) region. J. Farm Sci.,31(4): (452-456).

Ghanghas (2018). "economical status of farmers of vadodara in gujrat" J. Indian Soc. Coastal agric. Res. 30(1): 79-83

Gupta, S.C. (2018). A Critical Appraisal of PradhanMantriFasalBimaYojana (With Special Reference to Rabi Fasal 2016-17 in Rajasthan State).Inakshi publication.VOL-3 I-9 P: ISSN No.: 2394-0344.J KrishiVigyan 2019, 7 (2) : 46-50

Kiran, S and Shenoy, S.S. (2010). Constraints in adoption of system of rice intensification in Warangal district of Andhra Pradesh. Journal Research ANGRAU 38 (1 & 2):77-85.

Kumar, G.A.K. and Washon, M. (2007). Improving rice productivity in Orissa : A case of rice development scheme. Indian J. Ext. Edu. 43 (1&2): 49-55.

Mahesh, P;Bhanuprakash, M and Nirajkumar, S. (2011). Farmers empowerment through participatory on farm trials in rainfed rice ecosystem of Koderma, Jharkhand. Indian Journal Extension Education 9 (2):23-29.

Man, N and Sadiya, S.I. (2009). Off-farm employment participation among paddy farmers in the muda agricultural development authority and kemasinsemerak granary areas of Malaysia. Asia Pacific Development Journal. 16 (2): 141-153.

Mandal P.K., Jirli B. (2018) "social status of mustard farmers of Muraina district of M.P." Agric. Update, 7(3&4) Aug. 162-165

Nirmala, K. (2012). A study on diffusion status and adoption of System of Rice Intensification (SRI) in Mahaboobnagar district of Andhra Pradesh. M. Sc. (Ag.) Thesis. Acharya N.G. Ranga Agricultural University, Hyderabad, India.

Obaiah, M. C. (2004). A study on capacity building of rice growing farmers of Farmers Field Schools (FFS) in Krishna - Godavari zone of Andhra Pradesh. Ph.D. Thesis. Acharya N.G. Ranga Agricultural University, Hyderabad, India.

Painkra, S.K., Dev, C.M and Mandal, B.K. (2010). Information sources of tribal rice growers of Bastar district of Chhattisgarh. Journal CommunicationStudies. 28 (9):135-139.

Prasad, R.C.C. (2002). A study on the impact of On-Farm Extension Demonstrations (OFEDs) in Rice in Nellore district of Andhra Pradesh. M.Sc.(Ag.) Thesis. ANGRAU, Hyderabad, India.

Rai and Rai (2012) "Socio economic conditions of farmers in Udaipur" Raj. J. Extn.71 Edu.19 : 144-147

Singh Vikram and Ramchandran (2019) "study on the socio-economic status of the respondents gram panchayat in rural development in district Kanpur dehat (Uttar Pradesh), India Plant Archives Vol. 17 No. 1, pp. 723-726

11

Association Between Profile of Rice Grower and Their Extent of Adoption of Rice Production Technology by the Farmers of North Gujarat

Chandni Patel[1] and G. R. Patel[2]

[1]Department of Agricultural Extension and Communication, C.P. College of Agriculture, Sardarkrushinagar Dantiwada Agricultural University Sardarkrushinagar, Gujarat

[2](Academic), Sardarkrushinagar Dantiwada Agricultural University Sardarkrushinagar, Gujarat

Introduction

Among the rice growing countries, India has the largest area (448 lakh ha) followed by china and Bangladesh. Total food grain production in India is estimated at 28131.7 lakh tonnes. Rice production is estimated at 1156 lakh tonnes (Anonymous, 2010). In India, rice is grown in the state of Kerala, Andhra Pradesh, Orissa, Bihar, Uttar Pradesh, Madhya Pradesh, Punjab and West Bengal lead in the area, while West Bengal harvest highest rice production. The average yield per hectare is highest i.e., 5970 kg/ha in the Punjab during 2018-19.

Rice is grown throughout the state except Kachchh and Saurashtra. Some part of the state is growing rice as a drilled, while transplanted rice is taken in south and middle Gujarat. In Gujarat rice occupies about 10.61 per cent of the gross cropped area of the state and accounts for around 25.5 per cent of the food grain production. It is grown on an average about 7.5 to 8.5 lakh hector of land comprising nearly 70 to 80 per cent low land (transplanted) and 15 to 20 per cent of upland (drilled) rice (Anonymous, 2020). During the 2018-19, the yield of kharif rice in Gujarat was around 2247.75 kg/ha. Yield of summer rice was 3130.96 kg/ha and total production of rice in the year of 2018-19 was 1911 tonnes (Anonymous, 2020). During the 2019-20, the yield of kharif rice was 2117.81 kg/ha and yield of summer paddy was 2996.36 kg/ha and total production of rice in the year of 2019-20 was 1929.98 tonnes (Anonymous, 2020).

In Gujarat some districts like Chhotaudepur, Tapi, Navsari have highest rice productivity which covered 1.279 lakh hector area with production and productivity respectively of 3.338 lakh tonne and 2582 kg/ha and some district like Mehsana, Narmada, Panchmahal, North Gujarat etc. have medium to low productivity of paddy. But in some part of North Gujarat like Gandhinagar and Sabarkantha have low productivity of rice.

North Gujarat is one of the important ricegrowing districts in the Gujarat state. Gandhinagar and Sabarkantha is one of the major rice growing districts in North Gujarat. Majority of the farmers in the area has undertaken the cultivation of paddy vary long time. Therefore, it is high time to assess the relationship between profile of rice growers and their extent of adoption of recommended paddy production technology by the farmer specifically of Sabarkantha and Gandhinagar district.

Objectives

1) To ascertain the association between the profile of rice growers and their extent of adoption of rice production technology

Methodology

The present study was confirmed to “Ex-post facto” research design. Multistage random sampling technique was used for selection of talukas, villages and respondents from Gandhinagar and Sabarkantha district. Two talukas from each district and four villages from each taluka and ten rice growers from each village were selected making a sample of 160 respondents. The data were collected using pre tested interview schedule and analysed by appropriate statistical tools.

The dependant variables were measured with the help of the scale and indices developed by the past researchers as well as structured schedules which were framed for purpose.

Results and Discussion

The main purpose of the present study was to study the relationship between profile of rice growers and their extent of adoption of recommended paddy production technology by the farmers specifically of Sabarkantha and Gandhinagar district.

Association between the profile of rice growers and their extent of adoption of rice production technology

Table 1: Association between the profile of the rice growers and their extent of adoption of rice production technology (n = 160)

Sr. No.	Independent variables		Correlation coefficient ('r' value)
i	**Personal variables:**		
	(1)	Age	-0.060^{NS}
	(2)	Education	0.036^{NS}
ii	**Socio-economic variables:**		
	(1)	Size of land holding	0.067^{NS}
	(2)	Annual income	0.0084^{NS}
	(3)	Social participation	0.247**
iii	**Situational variables:**		
	(1)	Source of irrigation	0.070^{NS}
	(2)	Yield gap	-0.022^{NS}
iv	**Communicational variables:**		
	(1)	Extension participation	0.208**
	(2)	Source of information	0.179*

NS= Non-Significant, **Highly Significant (5%), S*Significant(1%).

Table 1 indicated that out of nine independent variable four variables *viz.,* education, size of land holding, annual income and source of irrigation had positive, but non-significant correlated with extent of adoption of rice production technologies. Age and yield gap had negatively, but non-significant correlation. Social participation and extension participation had positive and highly correlation with extent of adoption of rice production technologies. Remaining variable source of information had positive and significant association with extent of adoption of rice production technology.

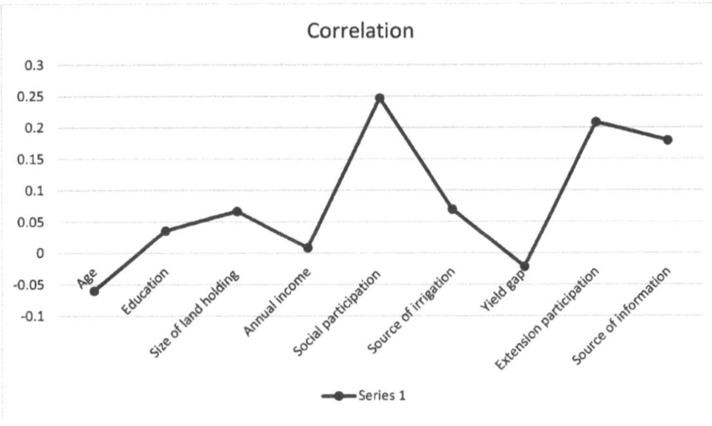

Fig. 1: Association between the profile of rice growers and their extent of adoption of rice production technology

Conclusion

The independent variables *viz.,* age and yield gap had negative and non-significant relationship with adoption of rice production technology. Education, size of land holding, annual income and source of irrigation had positive, but non-significant relation with adoption of rice production technology and social participation and extension participation had positive and highly significant relation with adoption and source of information had positive and significant relation with adoption of rice production technology

References

Anonymous (2010). Hand Book of Agriculture, ICAR, New Delhi. p. 817.

Anonymous (2020). State Agriculture Plan and State Infrastructure Development Plan (SSP and SIDP) (2017-18 to 2019- 20) Gujarat.

Vinaya Kumar, H. M., Shivamurthy, M., and Lunagaria, M. M. (2017). Impact of rainfall variability and trend on rice yield in Coastal Karnataka. Journal of *Agrometeorology.* 19 (3): 286-287.

12

Adoption Level with Respect to Various Practices of Mango Production Technology in Meerut District of Uttar Pradesh, India

Gaurav Kumar[1], R.N Yadav[2] , R. K Doharey[3], Ajay Pratap Singh[4] Arvind Kumar[5] and Dheerendra Kumar[6]

[1,3,4,5]Department of Extension Education, NDUA&T Kumarganj, Ayodhya Uttar Pradesh

[2,6]Department of Agricultural Extension & Communication, SVPUA&T Modipuram, Meerut, Uttar Pradesh

Abstract

The study was conducted in Meerut district of Uttar Pradesh to assess the present status of mango orchardists with the objective adoption level with respect to various practices of mango production technology. The Meerut district has comprises 12 community development block. Which two blocks were purposely selected according to need and availability of mango orchardists. On the basis of maximum area and production. 10 orchardists were selected from each selected village randomly with the help of progressive orchardists and village pradhan, thus the total sample size of 80 orchardists were selected for the present investigation. Appropriate adoption of the recommended new and improved mango production technology is a prerequisite for adopting any new innovation. The most of the respondents were having partially adopted about the improved varieties of mango, total sample size 61.25 per cent respondents. Maximum numbers of orchardists 45.00 per cent were observed in the partially adopted category the land preparation practise. Orchardists 56.25 per cent were observed in the not adopted category about the plantation practices. The not adopted category about the use of recommended dose of manures and fertilizer in mango orchard with 51.25 per cent. Maximum number of orchardists 58.75 per cent were observed in the partially adopted category in irrigation practises.

Keyword: *Mango practices, Technology practices .*

Introduction

The most significant commercially grown fruit crop in the country is the mango (*Mangifera indica* L.), which belongs to the Anacardiaceae family. It's known as the "king of fruits." Mango cultivars can be found in the greatest variety in India. Mango farming is thought to have begun in South East Asia. Since about 6,000 years ago, mango has been grown in south Asia.Mango is native to South Asia, from where it has been distributed worldwide to become one of the most cultivated fruits in the tropics.

The major mango-growing states are Andhra Pradesh, Uttar Pradesh, Karnataka, Bihar, Gujarat and Tamil Nadu. Andhra Pradesh rank 1st in area, whereas Uttar Pradesh rank 1st in production and productivity. Andhra Pradesh, Uttar Pradesh, Bihar, Karnataka, Maharashtra, West Bengal and Gujarat together contribute for about 82% of the total production in India.

India to become the second largest fruit producer in the world, after china, with a production of 102481 thousands metric tonnes of fruits from an area of 6930 thousands hectares in 2020-21.The fruit is grown in India, the total area under mango cultivation was 2317 thousand hectare with an annual production of 20386 thousand metric tonnes.(NHB 2020-21).

Mango is well adopted in tropical and sub-tropical climates but it can be grown up to 1,100 meters above sea-level. There should not be humidity, rain or frost during flowering. The temperature between 24 and 27^{0}C is ideal for its cultivation. Higher temperature during fruit development and maturity gives better quality fruits. The areas expressed frequent showers and high humidity is prone to many pests and diseases. Thus it can grow best in regions with a rainfall between 25cm and 250 cm. Regions having bright sunny days and moderate humidity during flowering is ideal for mango growing.The farmers 92.20 per cent had problems of pests in their mango trees. The most common pests were the mango seed weevil 48.10 per cent, fruit fly 36.40 per cent, trips 1.30 per cent, aphids 1.30 per cent and the most common diseases were the powdery mildew 46.80 per cent, anthracnose 18.20 per cent and scab 2.60 per cent (Gitonga,2010).

It is a good source of vital protective nutrients like vitamins such as vitamin A (1400 I.U.), and C. Mango fruit contains 73.0-86.7 per cent moisture, 11.6-24.3 per cent carbohydrate, 0.3-1.0 per cent protein, 0.1-0.8 per cent fat, 0.3-0.7 per cent mineral, 650-25900 μg vitamin 'A' and 3-83 mg vitamin 'C' per 100 gram fruit. Seed kernels contain 9.5 per cent protein, 8-12 per cent fat, 79.2 per cent starch, 2 per cent mineral matter and 2 per cent fibres (Tanwar, 2013).

It is a mango orchardists 57.50 per cent have full adoption of 10 m x 10 m planting distance. Most of the orchardists have full adoption of Dashehari, Langra and Chousa varieties for plantation in mango orchard. In fertilizers and manure application, most of the mango growers were have full adoption. Most of orchardists have partial adoption of ring and furrow method for irrigation, while 67.50 orchardists have full adoption of flood method for irrigation, There was 100 per cent full adoption in wettable sulphur, 97.50 per cent karathan and hexaconazal for disease control. Most of the orchardists have full adoption in Monocrtophos, Endosulphan and Imidacloprid for pest management. Maximum farmers were having full adoption in grading and packing of mango, while few orchardists have partial adoption in processing practices in mango. The 100 per cent mango orchardists have full adoption in Mandy samities for marketing of fruits, while most of the orchardists have no adoption of direct sale of fruits to consumers and whole sellers in same town and village and export of fruits (Tanwar et al., 2013).

Materials and Methods

The present study was conducted in Meerut district of Uttar Pradesh because the area under mango cultivation is adequate in this district. The Meerut district has comprises 12 community development block which two blocks namely Sardhana and Machhara were purposely selected according to need and availability of mango orchardists. The revenue villages was arrange in descending order based on the maximum area and maximum number of mango Orchardists, top 4 revenue villages were select from the each block. On the basis of maximum area and production. 10 orchardists were selected from each selected village randomly with the help of progressive orchardists and village pradhan, thus the total sample size of 80 orchardists were selected for the present investigation. Data were collected with the help of pre structured interview schedule covering all aspect of the present study. To make the procedure and information reliable and accurate, the investigator himself collected data with every individual mango orchardists either at his farm or his home. Scheduled developed was used for collecting information comprehensively. Before collecting information the purpose of the interview and study as a whole were explained to the orchardists. Methodologies of tools are discussed below

1. Percentage

$$\text{Percentage} = \frac{\text{Frequency}}{\text{Number of respondent}} x 100$$

2. Mean (X̅) $\sigma = \sqrt{\frac{1}{n}\sum_{i=1}^{N}(x_i - \overline{x})^2}$

Where,

(X̅) = Average or mean

$\sum$ x = Total number of scores obtained by respondents

N = Total number of respondent

3. **Standard Deviation**

$$Correl(X,Y) = \frac{\sum(x-\overline{x})(y-\overline{y})}{n}$$

Where,

σ = Standard deviation

d = Deviation from variables mean

n = Total number of items

Result and Discussion

To assess the Adoption Level with Respect to Various Practices of Mango Production Technology

Mango plays an important role in fruit cultivation scenario of Western Uttar Pradesh. It is known as the "king of fruits" many researchers are going to increase the productivity and total production of mango. As of result of them various new production technologies regarding mango orchard are been evolved. In spite of this, there has not been much increase in production as well as in its productivity. It is due to low adoption rate of mango production technology by the mango orchardists.

Table 1: To assess the adoption level with respect to various practices of mango production technology.

S. No.	Statements	Fully adopted		Partially adopted		Not adopted	
		F	P	F	P	F	P
1	Improved varieties	17	21.25	49	61.25	14	17.50
2	Land preparation	14	17.50	36	45.00	30	37.50
3	Plantation practices	4	5.00	31	38.75	45	56.25
4	Manures and fertilizers application	4	5.00	35	43.75	41	51.25
5	Irrigation practices	7	8.75	47	58.75	26	32.50

S. No.	Statements	Fully adopted		Partially adopted		Not adopted	
		F	P	F	P	F	P
6	Use of plant growth regulators	1	1.25	16	20.00	63	78.75
7	Plant protection measures	9	11.25	33	41.25	38	47.50
8	Suitable inter crops in mango orchards	10	12.50	50	62.50	20	25.00
9	Storage of fruit after harvesting	8	10.00	37	46.25	35	43.75
10	Packing Procedure (for long and short distance)	13	16.25	27	33.75	40	50.00
11	Marketing procedure	39	48.75	21	26.25	20	25.00

F-frequency, P-percentage

Improved Varieties

The data from the table 1 revealed that the most of the respondents were having partially adopted about the improved varieties of mango. Among the total sample size 61.25 per cent respondents were reported under partially adopted category, 21.25 per cent of respondents were reported under fully adopted category about the improved varieties of mango. Only 17.50 per cent respondents were founded under not adopted category about the improved varieties of mango.

It was concluded with the help of descriptive data of block Saradhana and machhara that the Dashehari, Langra, Chausa, Fazli, Ramkelaetc, were the more commonly used varieties of mango. The orchardists of the selected villages reported that they did not replace the tree unless and until it death, when the new tree was planted the attention was given on the new varieties. But only those varieties were planted which had already been tested for its productivity.

Land Preparation

The data from the table 1 revealed that the most of the respondents were having partially adopted about the land preparation practices of mango orchard. Among the total sample size 45.00 per cent respondents were reported under partially adopted category, 37.50 per cent of respondents were reported under not adopted category about the land preparation practices of mango orchard. Only 17.50 per cent respondents were under fully adopted category about the land preparation practices of mango orchard.

Plantation Practices

The data from the table1 showed that the most of the orchardists were not adopted of adoption about the plantation practices of mango plant. Among the total sample size 56.25 per cent orchardists were reported under not adopted appropriate plantation practices, 38.75 per cent orchardists were reported under partially adopted about the plantation practices of mango plant. Only 05.00 per cent orchardists were reported under fully adopted about the Plantation practices of mango plant.

Manures and Fertilizers Application

The data from the table1 showed that the most of the orchardists were not adopted about the use of recommended dose of manures and fertilizers in mango orchard. Among the total sample size 51.25 per cent orchardists were reported under not adopted category, 43.75 per cent orchardists were reported under partially adopted about the use of recommended dose of manures and fertilizers in mango orchard. Only 05.00 per cent orchardists were reported under fully adopted about the use of recommended dose of manures and fertilizers in mango orchard.

Irrigation Practices

The data from the table1 observed that the most of the orchardists were **partially adopted** about the irrigation practices of mango orchard. Among the total sample size 58.75 per cent orchardists were reported under partially adopted category, 32.50 per cent orchardists were reported under not adopted category about the irrigation practices of mango orchard. Only 08.75 per cent orchardists were reported under fully adopted category about the irrigation practices of mango orchard.

Use of Plant Growth Regulators

The data from the table1 revealed that the most of the orchardists were not adopted about the use of plant growth regulator for proper growth of mango fruits plants. Among the total sample size 78.75 per cent orchardists were reported under not adopted category, 20.00 per cent orchardists were reported under partially adopted category about the use of plant growth regulator for proper growth of mango fruits plants. Only 01.25 per cent orchardists were reported under fully adopted about the use of plant growth regulator for proper growth of mango fruits plants.

Plant Protection Measures

The data from the table1 showed that the most of the orchardists were not adopted about the proper plant protection measures in mango orchard. Among the total sample size 47.50 per cent orchardists were reported under not adopted category, 41.25 per cent orchardists were reported under partially adopted category about the plant protection measures in mango orchard. Only 11.25 per cent orchardists were reported under fully adopted category about plant protection measures in mango orchard.

Suitable Inter Crops in Mango Orchards

The data of the table1 showed that most of the orchardists were **partially adopted** about the suitable inter crops in mango orchards. Among the total sample size 62.50 per cent orchardists were reported under partially adopted category, 25.00 per cent orchardists were reported under not adopted category about the suitable inter crops in mango orchard and remaining 12.50 per cent orchardists were reported under fully adopted category about the suitable inter crops in mango orchards.

Storage of Fruit after Harvesting

The data from the table1 showed that the most of the orchardists were partially adopted about the correct storage method of fruits after harvesting. Among the total sample size 46.25 per cent orchardists were reported under partially adopted, 43.75 per cent orchardists were reported under not adopted about the correct storage method of fruits after harvesting. Only 10.00 per cent orchardists were reported under fully adopted category about the correct storage method of fruits after harvesting.

There is a more technological gap looks in storage of fruits after harvesting and its need to solve this problem by provide suitable extension strategy to the orchardists in proper time.

Packaging Procedure (for long and short distance) Market

The data from the table1 represent that the most of the orchardists were not adopted about the proper packaging methods of mango fruits for disposal at short and long distance market. Among the total sample size 50.00 per cent orchardists were reported under not adopted category, 33.75 per cent orchardists were reported under partially adopted category about the proper packaging of mango fruits for disposal at short and long distance market and remaining 16.25 percent orchardists were reported under fully adopted category about the proper packaging of mango fruits for disposal at short and long distance markets.

Marketing Procedure

The data from the table1 showed that the most of the orchardists werefully adopted about the marketing procedure of mango fruits. Among the total sample size 48.75 per cent orchardists were reported under fully adopted category, 26.25 per cent orchardists were reported under partially adopted category about the marketing procedure of mango fruits remaining 25.00 per cent orchardists were reported under not adopted category about the marketing of mango fruits.

Conclusion

From the above discussion we highlight the fact that the majority of the orchardists 61.25 percent were observed in the partially adopted category about improved varieties of mango. It was observed that the maximum numbers of orchardists 45.00 per cent were observed in the partially adopted category about the land preparation practise of mango orchard. Orchardists 56.25 percent were observed in the not adopted category about the plantation practices of mango plant. It was observed that the most of the orchardists 51.25 per cent were possessed in the not adopted category about the use of recommended dose of manures and fertilizer in mango orchard. It was clear from the study that the maximum number of orchardists 58.75 per cent were observed in the partially adopted category about irrigation practices in mango orchards. The maximum number of orchardists 47.50 percent were observed in the not adopted category about the proper plant protection measure in mango orchard. the maximum number of orchardists 62.50 percent were observed in the partially adoptedcategoryabout suitable inter crops in mangoorchards.the majority of the orchardists 78.75 percent were possessed in the not adopted category about use of plant growth regulatorsforproper of plant growth in mangoorchards.orchardists 46.25 percent were observed in the partially adopted category about the storage correct of method fruit after harvesting. the maximum number of orchardists 50.00 percent were possessed in the not adoptedcategoryabout packaging of mango fruits for disposal at long and short distance market respectively.It was observed that the maximum number of orchardists 48.75 percent were in the fully adoptedcategoryabout marketing procedure of mango.

References

Jadhav, B. (2009). Technological gap in adoption of recommended practices of mangocultivation. Master's thesis submitted toUniversity of Agricultural Sciences, Dharwad.

Kota, S. K. (2011). Knowledge and adoption of export oriented practices followed by the mango growers. Master's thesis submitted to Mahatma Phule Krishi Vidyapeeth, Rahuri.

Mahadik, R. P., Mehta, P. G. and Patil, V. G. (2008) Adoption of recommended mango cultivation technology by mango growers. Karnataka Journal of Agricultural Sciences; 21(2):314-315.

Mehta, B.M.andSonawane, M. (2012). Characteristic and adoption behaviour of mango growers in Valsad district of Gujarat. Agriculture Update;7(1/2):37-41.

Mirani, Z. (2007). Study of diffusion and adoption of Male Annihilation Technique. International Journal of Education and Development using Information and Communication Technology; 3(2): 89-99.

Modi, R. K.,Bheemappa, A., Manjunath, L., Hegde, R. V. and Handigol, J. A. (2010) Determinants of knowledge and adoption of post-harvest management practices among mango growers of northern Karnataka. Karnataka Journal of Agricultural Sciences; 23(4):606-611.

Moulasab, I., Jahagirdar, K. A. and Hirevenkagoudar, L. V. (2005). A study on adoption level of improved cultivation practices by mango growers of North Karnataka. Karnataka Journal of Agricultural Sciences;18(4):1143-1145.

Phuse, A. P., Vitonde, A. K. and Thipse, C.D. (2007). Adoption of recommended mandarin orange production practices. Indian Research Journal of Extension Education; 7(2&3): 98- 100.

Sharma, R. N., Singh, P. R. and Tiwari, O. K. (2011). Impact analysis of mango production training programme of Krishi Vigyan Kendra of Chhatisgarh. Agriculture Update;6(1): 88-91.

Shojaei, S. H., Hosseini, S. J. F., Mirdamadi, M. and Zamanizadeh, H. R. (2013). Investigating barriers to adoption of integrated pest management technologies in Iran.Annals of Biological Researc;4(1): 39-42.

Singh, K. V., Singh, G. P. and Priyadarshi, A. (2010). Extent of adoption of improved practices of mango production by mango growers in Muzaffarnagar district of Uttar Pradesh. Indian Research Journal of Extension Education; 10 (3):107-113.

Sumathi, P. and Rathakrishnan, T. (2008). Adoption of improved practices in mango cultivation by small and big farmers in Dharmapuri district. Madras Agricultural Journal; 95(1-6): 239-241.

Tanwar, S. R., Singh. D., Yadav, R. N., Singh, D. K. and Singh, V. K. (2013).Studieson adoption level of orchardists to package of practices for mango production. Annals of Horticulture; 6(1):99-106.

Thippeswamy, R. (2007). A study on knowledge and adoption of plant protection measures in coconut crop. Master's thesis submitted to University of Agricultural Sciences, Dharwad.

Thorat, K. S., Suryawanshi, D. B. and Ban, S. H. (2012). Technological gap in adoption of recommended cultivation practices of mango growers and constraints faced by them. Mysore Journal of Agricultural Sciences; 46(1):160-163.

Yadav, R. N. and Sharma, T. D. (2007). Knowledge and adoption level of orchardists regarding mango cultivation in Western Uttar Pradesh. Plant Archives; 7(1):389-393.

13

A Study on Adoption of Improved Brinjal Production Practices by the Farmers of East Champaran District of Bihar

***Sanskriti Sindhu*[1] and *Anurag Shankar Singh*[2]**

[1]*Krishi Vigyan Kendra, Parsauni, Bihar*

[2]*Extension Education, Acharya Narendra Deva University of Agriculture & Technology, Kumarganj Ayodhya, Uttar Pradesh*

Abstract

Vegetables has become as an essential requirement of the daily human diet, because of its nutritional value. Brinjal (Solanumme longena L.) or egg plant is one of the most common popular and principle vegetable crops grown in India and other parts of the world. The present study was conducted to understand the existing status of adoption of improved Brinjal production practices. This study was carried out in East Champaran district of Bihar during 2020-2021. Descriptive research design was adopted for the study. Byemploying random sampling technique, 120 brinjal growers were selected and data were collected by personal interview method by using pre-tested interview schedule and later appropriate statistical analysiswas done to find out the meaningful result. The results reported that majority of respondent's hadmedium level of adoption followed by low and high level of adoption on brinjal production technology among brinjal growers.

Keywords: *Adoption, Improved brinjal production practices, random sampling technique etc.*

Introduction

Horticulture crops cover large varieties of fruits, vegetables, flowers and plantation and spices crops. Among these, vegetable cultivation is the major attraction to farmers as it is comparatively more remunerative the field crops. Vegetables not only provide maximum output but also give more income per unit area of land. Vegetable cultivation among small land holders has always been source of supplementary source of income and provides gainful employment through intensive cultivation and thus vegetable growers are normally more prosperous than those who grow cereals, because of higher returns.

Brinjal (*Solanummelongena* L) is a commercial vegetable crop belongs to the family of solanaceae. It is a nannual crop. It is mainly used for culinary purpose. Brinjal has been considered as one of the important vegetable crop. The brinjal growers can increase production of brinjal through adoption of new varieties and with improved technologies. A first-hand knowledge regarding existing status of adoption of improved technology helpful to the extension workers for concentrating their efforts to create favorable condition for better adoption of the innovations of brinjal cultivation. Arvind kumar singh (2017) revealed that more than half of the respondents belong to middle age group, educated up to higher secondary, had medium level ofland holding, medium level of annual income, sources of information and low level of material possession. Further headded, that he possessed low level of farm power.

Bihar contributes share of 9.72% with production of 1241.71 tonnes. (National Horticulture Board: 2019-2020). It is widely grown vegetable of Asia, parts of Europe and Africa. It is one of the most traditional vegetables in India. Brinjal is grown extensively in India, Bangladesh, Pakistan, China and the Philippines.

Brinjal had high nutritive and medicinal values. It containsabout 92.3 gms moisture, 0.3gms fat,1.3 gms fiber, 24 kcal energy, 1.4 gms protein, 0.3 gms minerals, 4 gms carbohydrates, 12 mgsvitamin C etc. per 100 gm. Brinjal fruit (unripe) is primarily consumed as cooked vegetable invarious ways and dried shoots are used as fuel in rural areas. It is low in calories and fats, containmostly water, some protein, fibre and carbohydrates. It is a good source of minerals and vitamins and is rich in total water soluble sugars, free reducing sugars, amide proteins among othernutrients.

Research Methodology

Descriptive research design was adopted for the study since it was a scientific method which involves observing and describing the behaviour of the subject without influencing it in anyway. East Champaran district of Bihar was purposively selected for the present study. Since Brinjal isthe major crop cultivated in the area. In East Champaran district of Bihar, there are 27 blocks. Out of 27 blocks, Turkaulia block was selected purposively since it possessed the maximum area under brinjal cultivation. In Turkaulia block of East Champaran district of Bihar, there are 34 villages. Out of 34 villages, 6 villages were selected; namely, Bijulpur, Chargaha, Hardia,

Jaisinghpur, Madhopur and Turkauliya. From the selected six villages namely Bijulpur, Chargaha, Hardia, Jaisinghpur, Madhopur and Turkauliya, 120 brinjal growers were selected. The complete list of brinjal growers were collected from the agriculture department. From eachvillage, 20 brinjal growers were selected; thus, from six villages, 120 brinjal growers constitute the respondents. The primary data were collected with the help of interview schedule, which was prepared on the basis of the objectives of the study. The secondary data were obtained fromdepartment of agriculture, various Governments like District Agriculture office, Sub-divisionoffice, Block Development Office, and from magazines, publications, previous studies and other thesis.

For calculating percentage, frequency was multiplied by 100 and divided by total number of Respondents.

$P = X/N \times 100$

Where,

P = Percentage

X = Frequency of the Respondents

N = Totalnumber of Respondents

Mean: Mean was obtained by dividing the sum of thescores by the total number of cases involved. The formula for determining mean is

$Mean = \sum X/n$

Where,

$\sum X$= sum of

scoresn = no.of observation

$vation_{Mn}$

Standard Deviation

The standard deviation is the square root of the arithmetic average of the square of various from their arithmetic mean.

S.D.=

Where,

$\sum X$ = Deviation of the score from

meann = Number of observation

The responses were recorded, classified and tabulated and appropriate statistical tools were employed and the results were presented as follows.

Results and Discussion

Table 1: Distribution of Brinjal growers according to their adoption level (n=120)

S. no.	Recommended improved cultivationpractices	Response					
		Fully adopted		Partially adopted		Not adopted	
		f	%	*F*	%	*f*	%
1	Cultivation Practices	82	68.30	32	26.70	6	5.00
2	Nursery Sowingtime	60	50.00	48	40.00	12	10.00
3	Planting time a) Early b) Mid c) Late	62	51.70	30	25.00	28	23.30
4	Seed Treatment	58	48.33	37	30.84	25	20.83
5	Seedrate: a) 250-300g/ha b) 300-400 g/ha	79	65.83	26	21.67	12	12.50
6	Source of Seed Material: a) Privateshop b) Government Centre c) Agriculture Research Station	80	66.67	27	22.50	13	10.83
7	Method of Sowing: a) Broad Casting b) Transp Lanting	71	59.17	33	27.50	16	13.33
8	Spacing: 60× 45cm75×60 cm	82	68.33	27	22.50	11	9.17
9	Chemical Fertilizer and Manure Management: a) FYM b) Nitrogen c) Phosphorus d) Potassium	68	56.67	24	20.00	28	23.33
10	Intercultivation	71	59.17	33	27.50	16	13.33
11	Weed Management: a) Hand Weeding	48	40.00	45	37.50	27	22.50
12	Plant Protection Measures	82	68.33	27	22.50	11	9.17
13	Yield:- a) Early b) Medium c) Late	62	51.67	37	30.83	21	17.50

Majority of the respondents fully adopt the varieties for sowing (68.30%), followed by 26.70 percent partially adopted and only 5 per cent not adopted.

Half of the respondents adopted thenursery sowing time (50%), followed by 40 per cent partially adopted and 10 per cent notadopted. 51.70 per cent, 25 per cent and 23.30 per cent fully adopted, partially adopted and notadopted the planting time. Regarding seed treatment, 48.33 percent, 30.84 percent and 20.83 per cent were fully adopted, partially adopted and not adopted the cultivation practices. Morethan half of the respondents (65.8%). 66.67 per cent, 22.50 per cent and 10.83 per cent were fully adopted, partially adopted and not adopted the seed material source. More than half of the respondents (59.17%) were fully adopted, 27.50 per cent partially adopted and 13.33 per cent not adopted sowing methods.

Majority (68.33%, 22.50% and 9.17%) of respondents were fully adopted, partially adopted and not adopted any practices. Most of the respondents fully adopted (56.67%), partially adopted (20%) and not adopted (23.33%). Regarding inter cultivation, fully adopted, partially adopted and not adopted of cultivation practices were 59.17 percent, 27.50 percent and 13.33 per cent respectively. 37.50 per cent, 40 per cent and 22.50 per cent of respondents partially adopted, fully adopted and not adopted any cultivation practices. With respect to plant protection measures, 68.33 per cent fully adopted, 22.50 per cent partially adopted and 9.17 per cent not adopted any cultivation practices. Regarding yield, 51.67 percent of respondents fully adopted,

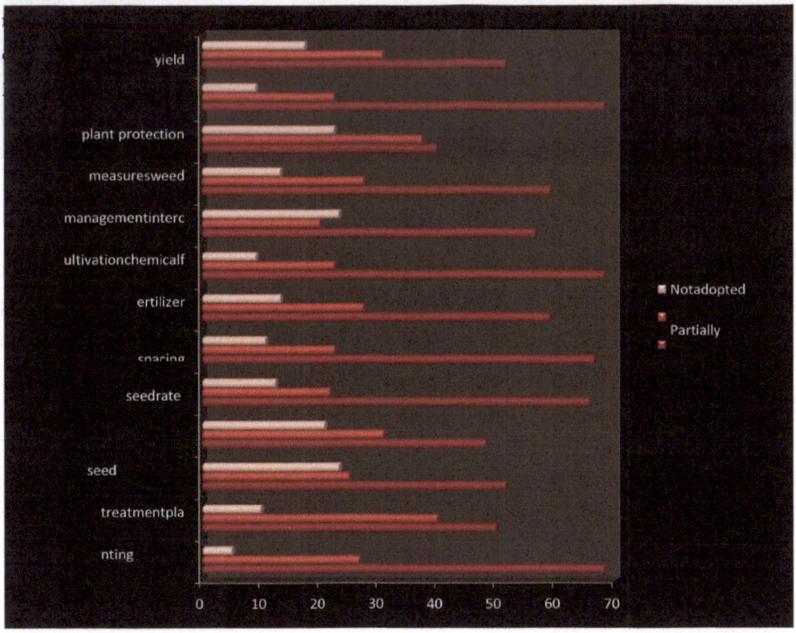

Fig. 1: Distribution of respondents according to their Adoption level of brinjal growers

Table 2: Distribution of brinjal grow ersaccordingtotheiradoption level

S.No.	Adoption	Frequency	Percentage
1.	Low(18-24)	40	33.34
2.	Medium(25-30)	56	46.66
3.	High(31-36)	24	20.00
	Total	120	100.00

Table 2. reveals that higher percentage of the respondents (46.66%) had medium level of adoption, followed by low (33.34%) and high (20%) level of adoption on brinjal production technology among brinjal growers.

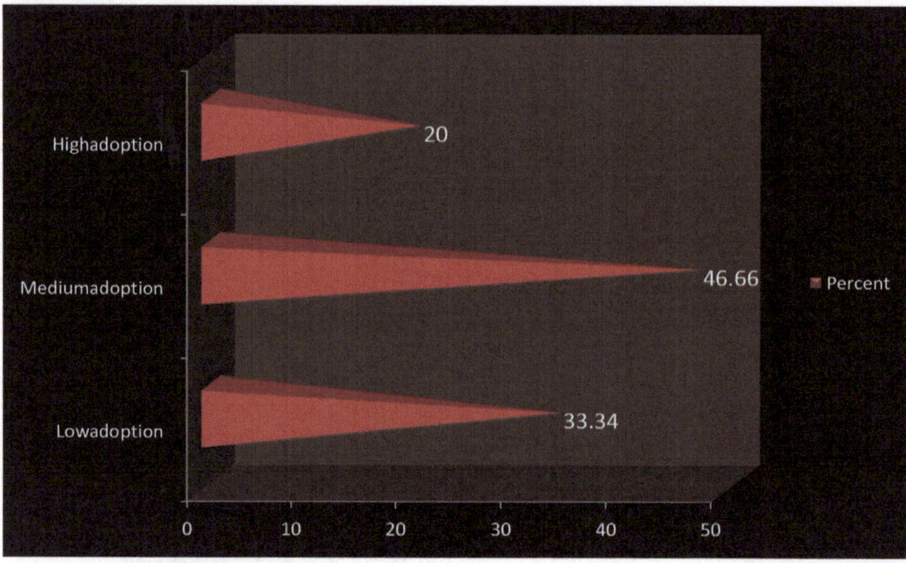

Fig. 2: Distributionofrespondentsaccordingto Overall Adoption

Conclusion

It was concluded that Brinjal (Solanummelongena L.) is cultivated as one of the leading vegetable crops grown in India. Proper efforts should be made to adopt improved production practices of brinjal. Government support and allocation of resources would lead to adoption of improved production technology. The present study concluded that majority of respondents had medium level of adoption followed by low and high level of adoption on brinjal production technology among brinjal growers.

References

Barodia, A.R. 2005. A study on adoption behaviour of vegetable growers in relation toscientific technology in Phanda block ofBhopal district. M.Sc. (Ag.) Thesis (Unpublished),JNKVV, Jabalpur.

Indian production of brinjal (2019-2020).Source: National Horticulture Board,http://www.nhb.gov.in.

Kushwaha, R.G. 2003. Behaviour of vegetable growers towards plant protection measures in Solanaciouscropsin Ghatigoasblock of Gwalkiordistrict, M.P.M.Sc., (Ag.) Thesis (Unpublished.), JNKVV, Jabalpur.

Lokhande, VInod Kumar. 2010. A study on adoption behaviour of tomato growers in relation to improved production technology in Chhindvarablock of Chhindvaradistrict. (M.P.) M.Sc.(Ag.) thesis (Unpublished), JNKVV, Jabalpur.

National Horticulture Board (2018-2019). Area and production of horticulture crops (3rd advances timates)

Shriwas Y. 2011. A study on knowledge and adoption of recommended brinjal production technology among the farmers of Durg district of Chattisgarh state. M.Sc., (Ag.) Thesis. IndiraGandhiKrishiVishwavidyalaya.

Shriwas,Y.,Sarkar,J.D.,Awsthi,H.K.andSarthi,N.(2015).KnowledgeofRecommended Brinjal Production Technology among the Farmers. Department of Agricultural Extension, Indira Gandhi Krishi Vishwavidyalaya, Raipur- 492 012 (C.G.), S.M.S., KrishiVigyan

Singh A.K. 2017. A study on knowledge and adoption of recommended brinjal production technology among the farmers of Gohadblok of Bhind district of Madhya Pradesh.M.Sc. (Ag.) Thesis RVSKVV, Gwalior.

Singh D. 2018. Impact of Front Line Demonstrations on Productivity of Brinjalcv. Nunhems BE-707 in Bharatpur District of Eastern Rajasthan, India. Int. J. Curr. Microbiol. App. Sci. 7(03): 1287-1291.

Walke.A.S., P.G. Khalache and J.H. Gaikwad. 2009. Study of knowledge and adoptionlevel of the brinjal growers about recommended technology and constraints in Rahuri tehsil of Ahmednagar district .Agriculture Update. Februaryto July. Vol.4(1&2):93-95.

14

Assessment of Livelihood Status of Trained Farmers of Himachal Pradesh

Neha Dhadwal, Kanika Mehta and Rajesh Kumar Thakur and Komal Sharma

Department of Agricultural Economics, Extension Education and Rural Sociology CSK Himachal Pradesh Krishi Vishvavidyalaya, Palampur Himachal Pradesh

Abstract

Agriculture is undoubtedly the major source of livelihood for majority of the rural population in the country. Especially in hilly states like Himachal Pradesh, it is an imperative source of income and livelihood security. The livelihood security mainly comprises of components like food, financial, health, cultural, educational securities. Factors like fragmented land holdings, illiteracy among farmers, lack of market access, low production adversely affect these securities, ultimately affecting the livelihood status of farm households. Among different districts of the state, district Kangra was selected for the study as majority of the farmers belonged to this district. The primary data were collected from 60 trained farmers through a two-stage sampling design. The present study is, therefore aimed to assess the livelihood security of farmers practising vegetable cultivation and protected cultivation in the state. The trained farm households were found to be more secured in terms of education (0.46) followed by economic security (0.30), food (0.27) and social security (0.20). The values ranged between 0.05 to 0.82 in case of education security, 0.03 to 0.85 in food security, 0.02 to 0.65 for economic security and 0.01 to 0.63 in case of social security index on overall farm situation. The farm category wise comparison revealed that vegetable farming category households were relatively more secured in case of educational and food securities while in case of economic and social security aspects the integrated farming households were relatively better. Various recommendations were also made based on the results to improve the livelihood security of trained farm households.

Keywords: *Livelihood, trainings, agriculture, farm income*

Introduction

Agriculture has been undoubtedly the major source of livelihood in the developing nations like India. Livelihood is defined as adequate stock and flow of food and cash with an individual or a family to meet its basic needs (Girish *et al.*, 2020). It encompasses the capabilities, assets (stores, resources, claims and access) and activities required for a means of living (Chambers and Conway, 1992). Livelihoods are considered secure when households have secure ownership or access to resources and income earning activities, including reserves and assets, to offset risks, ease shocks and meet contingencies (Chambers, 1989). The livelihood security is a complex concept that is location specific, subjective and dynamic (Hogger and Ruedi, 2004). It comprises of many indicators like food, educational, financial, health, culturaland social securities. Constraints like practices of traditional methods of cultivation, lack of access to modern technology, low productivity of crops, in-adequate capital formation and low investment, widespread illiteracy among farmers, etc., have become a major threat to the educational, food, economic and social securities, thus ultimately affecting the livelihood status of farming households in the country (Hatai and Sen, 2008). Since, a large fraction of farmers in the country lives in the rural areas, they have no proper knowledge regarding the usage of modern farming techniques. Often their low level of education and lack of communication has resulted into a lack of general awareness regarding the modern research and inventions on agriculture (Makal et al., 2017). In such scenario, farmers' trainings play important role which helps the farmers in understanding the application of latest agricultural technologies in daily farming activities. It improves the knowledge of the trainees about the improved farm practices (Srinivas and Sailaja, 2013). The major players providing these extension services in public sector are Krishi Vigyan Kendras (KVKs), State Agricultural Universities (SAUs) and ICT-led extension interventions by Department of Agriculture, Cooperation and Farmers Welfare (DAC&FW), Government of India. Many state universities through their KVKs, have been imparting need-based trainings to the farmers in their routine activities as well as sponsored by various departments/agencies. In last few decades, KVKs have increasingly become associated with grass root development (Saha, 2013).

Various studies in the past have investigated the importance of trainings in uplifting the livelihood status of farmers. Noor and Dola (2011), analysed the impact of trainings on perception and performance level of livestock farmers which indicated various positive impacts of trainings viz. increase in farm products, enhanced work quality, increased income, increased level of skill, knowledge and ability. Similar results were reported in studies by (Singh and

Singh, 2014) and Sharma and coworkers (2017) which indicated improved skills and knowledge among farmers as a result of trainings. A comparative analysis of participants and non- participant farmers under KVK trainings was carried out by Meena and co workers (2015). The study revealed participant farmers as better adopters of knowledge and production technology in comparison to non-participators. Several other studies have also reported the need of improvement in the trainings provided to farmers. Meena & Singh (2010) analysed the impact of training programmes imparted by Krishi Vigyan Kendras in Rajasthan. The results emphasized the need for strengthening the training programmes and more efforts on the part of KVKs for improving production. Similar results were reported by Bar and coworkers (2014). The trained farmers are expected to bebetter at adoption of improved agricultural technologies in their production systems due to which their livelihood status may have improved as reported by earlier studies. Therefore, it is imperative to assess the livelihood security of the trained farmers. The evidences from above literature help to understand the importance of livelihood security of farm households. Taking various aspects of above reviews into consideration, the present study was conducted to access the livelihood status of farmers of Himachal Pradesh.

Methodology

The present study was carried out in Kangra district where a large group of rural households had been trained in two broad areas namely- integrated farming and vegetable farming respectively. A total of 60 trained farm households were interviewed for the primary data collection which comprised of two categories i.e. integrated farming and vegetable farming (30 households each). The respondents were selected through two stage sampling technique. At the first stage of sampling, out of 6 major fields of specialization i.e. integrated farming, vegetable farming, mushroom cultivation, bee keeping, protected cultivation, organic farming two fields of specialization i.e. integrated farming and vegetable farming were selected randomly. At the second stage of sampling, a sample of 60 trainees (comprising 30 traineesfrom each selected category) was selected randomly from the secondary data obtained from the institution. The primary data were collected through pre-tested survey schedule by personal interview method.

Analytical Framework: The livelihood status of the trained farm households was computed using the Composite Livelihood Security Index (Venu et al., 2017 and Hatai and Sen, 2008). For calculating the CLSI, four major domains were taken into consideration i.e. economic, food, educational and social security.

The livelihood security index was analysed using a balanced weighted average approach where each domain is assumed to contribute equally to the overall index. These domains were further categorized into different components which are specified as below:

Educational security: It comprised of three components i.e.average years of schooling of family members, number of educational institutions accessible (within 2 km)and relative accessibility of educational institutes (distance in km) to trained households.

Food security: It consisted of components like per capita consumption of rice (kg per month), per capita consumption of wheat (k per month), per capita consumption of vegetables (kg per month) and per capita consumption of milk (litres per month) of trained farm households.

Economic security: It consists of four components namely value of farm assets excluding land (Rs.), on farm income (Rs.), off-farm income (Rs.) and landownership (in ha) of trained households.

Social security: It includes components like family members educated above senior secondary, active workers/ family, number of females in government jobs, females in private job and social participation (NGOs, Self Help Groups (SHGs), co-operative societies, etc.)

Since, each component has been measured on a different scale, therefore, the components have been standardized by using the approach adopted in measuring "Life Expectancy" in Human Development Reports (Shaheen and Sanzidur, 2012).

$$Correl(X,Y) = \frac{\sum(x-\bar{x})(y-\bar{y})}{\sqrt{\sum(x-\bar{x})(y-\bar{y})^2}}$$

where,

Z_{ijk}= standardized components of each domain

i = domains (i = 1, 2, 3 and 4 for educational, food, economic and social domains, respectively)

j = components (j = 1, 2, 3, ….., j)

k = farmers (k =1, 2, 3, ………, 60)

Once each component representing a particular livelihood security domain was standardized, the relevant livelihood security indices for each domain i.e. educational, food, economic and social were constructed by averaging their respective standardized components by using the formula given as below:

$$LSI_{ik} = \frac{\Sigma Z_{ijk}}{J}$$

where,

j = components (j = 1, 2, 3, ….., j)

k = farmers (k =1, 2, 3, ………, k)

J = no. of components used to construct a particular domain

The Composite Livelihood Security Index (CLSI) for the sample households was obtained using the formula:

$$CLSI_k = \frac{\Sigma W_{ik} LSI_{ik}}{N}$$

where,

Σ LSI_{ik}= Livelihood Security indices for each domain

N = Total number of domains

W_{ik} = Weight assigned to the ith domain of LSI

If the weights are identical and sum up to unity i.e. ($W_{i1} + W_{i2}+W_{i3+} W_{i4} = 1$) then CLSI is calculated as a simple mean. Butwhen the weights are different across all domains then CLSI is calculated as a weighted mean. For distinction, the former has been denoted simply as 'CLSI' and the latter as 'CLSI*'.

The CLSI* was calculated by firstly taking the inverse of the proportional contributions of Educational Security Index, Food Security Index, Economic Security Index and Social Security Index to CLSI is to be obtained. Then, the weights to be assigned to each domain will be the ratio of its inverse contribution to the sum of all the four inverse proportions. The value varies between 0 and 1 (Ghabru*et al.*, 2017). A value closer to 0 shows low livelihood security and value closer to 1 show high level of livelihood security.

Results and Discussion

The results and discussions pertaining to livelihood security of farm households have been discussed as below:

Educational security: The education helps the farmers in making rational decisions and is the key for securing better livelihood and profitability in farm business. Better is the education, the better well fed the people, the better would be capacity, capability and appreciation of human beings to be better productive economically (Singh, 2000). The details of various components of education security have been given in Table 1. The analysis revealed that 42 per cent of households had average years of schooling of 6-10 years followed by 12 per cent households having average years of schooling of 11-15 years. Moreover, 27 per cent households had middle and high schools in the vicinity of 2 km distance while 17 per cent households had senior secondary schools in the periphery of 2 kms. The results clearly indicated the better educational facility in study area.

Food security: The food security is a basic and essential component of livelihood security. The consumption pattern of food items vary from region to region depending on the food habits and average income of households. There are various types of food items consumed in a family like pulses, cereals, meat, fish, beverages, milk, eggs etc. Considering the food consumption pattern of study area, the consumption pattern of major food items by the sample households has been depicted in Table 2. The per capita consumption of the farm households has been depicted in Table 3.

Table 1: Average years of schooling, availability and accessibility of educational institutes to sample households

Sr. No.	Particulars	Integrated Farming	%	Vegetable Farming	%	Overall	%
1	**Average years of schooling**						
i.	0-5 years	5	16.67	-	-	5	8.33
ii.	6-10 years	23	76.67	19	63.34	42	70.00
iii.	11-15 years	2	6.66	10	33.33	12	20.00
iv.	> 15 years	-	-	1	3.33	1	1.67
2	**No. of institutions (within 2 km)**						
i.	Primary	8	26.67	5	16.67	13	21.67
ii.	Middle and high school	12	40.00	15	50.00	27	45.00
iii.	Senior secondary	9	30.00	8	26.67	17	28.33
iv.	Higher education	1	3.33	2	6.67	3	5.00
3	**Relative accessibility (distance in km)**						
i.	0-0.5km	8	26.67	6	20.00	14	23.33
ii.	0.6-1km	10	33.33	12	40.00	22	36.67
iii.	1.1-1.5 km	8	26.67	7	23.33	15	25.00
iv.	>1.5km	4	13.33	5	16.67	9	15.00

Table 2: Average monthly consumption of food items (kg/family/month)

Sr. No.	Particulars	Integrated Farming	Vegetable Farming	Overall
1	Rice	27.46	18.53	23.00
2	Wheat	26.60	21.33	23.97
3	Maize	7.50	5.82	6.66
4	Pulses	9.98	11.07	10.52
5	Fruits	4.00	4.25	4.13
6	Vegetables	16.80	14.30	15.55
7	Milk	25.48	28.70	27.09

Table 3: Per capita consumption of food items (kg/ month/person)

Sr. No.	Particulars	Integrated Farming	Vegetable Farming	Overall
1	Rice	5.35	3.71	4.54
2	Wheat	5.19	4.27	4.73
3	Maize	1.46	1.16	1.31
4	Pulses	1.95	2.21	2.07
5	Fruits	0.78	0.85	0.81
6	Vegetables	3.27	2.86	3.07
7	Milk	4.97	5.74	5.34

Economic security: The economic security of a household is reflected through its total annual earnings from various sources i.e. farm activities, off farm sources like government job, private services, land ownership, farm assets owned (value of all the major and minor farm implements owned by the farmer) etc. and has direct relationship with magnitude of annual income and livelihood security. The details of education security have been given in Table 4. About 23 per cent farm households were found to have farm assets worth more than Rs. 50,000. Majority of farm households (70 per cent) had annual farm income in the range of 2.5 lakh-5 lakhs. The ownership of land analysis shows that 58 per cent households had land holdings of size 0.4-0.6 ha.

Table 4: Range wise distribution of Economic security components of farm households

Sr. No.	Particulars	IF	%	VF	%	Overall	%
1	**Value of farm assets (Rs.)**						
i.	<10,000	22	73.33	22	73.34	44	73.33
ii.	10,000-50,000	1	3.33	1	3.33	2	3.33
iii.	>50,000	7	23.33	7	23.33	14	23.33
2	**Farm income**						
i.	<2.5 lakh	6	20.00	10	33.33	16	26.67
ii.	2.5 lakh - 5 lakh	23	76.67	19	63.34	42	70.00
iii.	>5 lakh	1	3.33	1	3.33	2	3.33
3	**Off farm income**						
i.	<2.5 lakh	24	80.00	24	80.00	48	80.00
ii.	2.5 lakh - 5 lakh	2	6.67	3	10.00	5	8.33

Sr. No.	Particulars	IF	%	VF	%	Overall	%
iii.	>5 lakh	4	13.33	3	10.00	7	11.67
4	Land ownership						
i.	<0.2 ha	3	10.00	7	23.33	10	16.67
ii.	0.2-0.4 ha	15	50.00	20	66.67	35	58.33
iii.	0.4-0.6 ha	11	36.67	1	3.33	12	20.00
iv.	>0.6 ha	1	3.33	2	6.67	3	5.00
	Total	30	100.00	30	100.00	60	100.00

Note: IF-Integrated farming, VF- Vegetables farming

Social security: Social security parameters hold equal importance like other parameters in assessing the livelihood security of farm households. The detailed analysis of social security indicators like education above senior secondary, active workers/family, females in government job and private job, social participation of farm households etc., have been presented in Table 5. The table shows that 15.46 per cent farm household members had attained qualification higher than senior secondary.In context of female employment, only 3.42 and 6.85 per cent females were found to be engaged in government jobs and private services, respectively. The social participation of farm households i.e. in Non-Governmental Organizations(NGOs), Self Help Groups (SHGs), panchayats, etc. was found to be 7.89 per cent.

Table 5: Social participation of sample households

Sr. No.	Particulars	IF	%	VF	%	Overall	%
1	Education above senior secondary	12	7.79	35	23.33	47	15.46
2	Active workers/family	0.61		0.53		1.15	
3	Females in govt. job	2	2.74	3	4.11	5	3.42
4	Females in private job	6	8.22	4	5.48	10	6.85
5	Social participation	13	8.44	11	7.33	24	7.89

Note: IF-Integrated farming, VF- Vegetables farming

Table 6: Educational, food, economic and social security indices for livelihood security index

Educational security index	Integrated Farming	Vegetables Farming	Overall
Average years of schooling	0.36	0.39	0.38
Quality of institutions (within 2 km)	0.56	0.55	0.55
Relative accessibility	0.42	0.46	0.44
Food security index(kg/month)			
Per capita consumption of rice	0.23	0.37	0.30
Per capita consumption of wheat	0.31	0.38	0.34
Per capita consumption of vegetables	0.23	0.24	0.26
Per capita consumption of milk	0.38	0.20	0.17
Economic security			

Educational security index	**Integrated Farming**	**Vegetables Farming**	**Overall**
Value of farm assets	0.15	0.17	0.14
On farm income	0.58	0.45	0.53
Off-farm income	0.36	0.29	0.30
Landownership	0.25	0.20	0.23
Social security			
Education above senior secondary	0.13	0.23	0.20
Active workers/family	0.55	0.42	0.50
Females in government jobs	0.07	0.05	0.04
Females in private job	0.10	0.13	0.08
Social participation	0.22	0.18	0.20

The individual indices for education security, food security, economic security, social security were computed to determine the livelihood security index as shown in Table 6. The values of composite livelihood security index have been presented in Table 7. The values of the major domains of livelihood security domains i.e. education, food, economic and social were used to worked out the overall livelihood security (LSI) and weighted livelihood security (LSI*) indices. The values of LSI and LSI* for the sample households were estimated as 0.31 and 0.39, respectively.

It is very important to determine the individual components of livelihood security since it is affected by many factors like education, social participation, consumption habits etc. The livelihood security has been operationalized as adequate access to income and other resources to meet the basic needs including food and nutrition, health, facilities, educational facilities and social integration (Barela *et al.*, 2018).The entire range of livelihood security of sample households in study area has been presented in Table 8.

Table 7: Computation of composite livelihood security index (CLSI) of sample households

Particulars	**Integrated Farming**	**Vegetables Farming**	**Overall**
Education security index	0.45	0.47	0.46
Food security index	0.28	0.30	0.27
Economic security index	0.34	0.28	0.30
Social security index	0.23	0.22	0.20
Overall Composite livelihood security index	0.32	0.31	0.31
CLSI *	0.40	0.40	0.39

Table 8: Range of livelihood securities of sample households

Particulars	Integrated Farming		Vegetables Farming		Overall	
	Low	High	Low	High	Low	High
Education security index	0.07	0.77	0.12	0.8	0.05	0.82
Food security index	0.07	0.89	0.04	1	0.03	0.85
Economic security index	0.02	0.64	0.04	0.67	0.02	0.65
Social security index	0.03	0.45	0	0.56	0.01	0.63
Overall livelihood security index	0.16	0.55	0.16	0.59	0.17	0.55
LSI *	0.24	0.64	0.21	0.73	0.22	0.66

Conclusion

The trained farm households were found to be more secured in terms of education (0.46) followed by economic security (0.30), food (0.27) and social security (0.20). The values ranged between 0.05 to 0.82 in case of education security, 0.03 to 0.85 in food security, 0.02 to 0.65 for economic security and 0.01 to 0.63 in case of social security index on overall farm situation. The farm category wise comparison revealed that vegetable farming category households were relatively more secured in case of educational and food securities while in case of economic and social security aspects the integrated farming households were relatively better. However, the overall livelihood security of farmers has not improved much. In order to address the problem of low land holdings and high cost of heavy machinery, it is suggested the farmer may be motivated to form the groups at the village level. The department of agriculture should provide tractor/power tiller, thresher, nursery planter, etc. to the group under ongoing schemes of the department in the state. There is a need for regular upgradation of skills of farmers by the officers of department of agriculture and subject matter specialists of Krishi Vigyan Kendras. In order to improve the extent of adoption of farm technologies, the timely availability of critical inputs especially seeds of high yielding varieties, plant protection material etc. should be ensured.

References

Barela HR, Jha SK, Rai CK and Yadav R. 2018. Assessment of livelihood security of tribal farmers: A case study from tribal areas of Madhya Pradesh, India. International Journal of Current Microbiology and Applied Sciences 7: 1135-1141

Bar N, Dash SR and Mishra P. 2014. Impact of KVK training on development of tribal farmers- A study in Odisha. Journal of Environment and Ecology 5

Chambers R. 1989. Editorial introduction: Vulnerability, coping and policy. IDS Bull 2(2):1-7

Chambers R and Conway G. 1992. Sustainable rural livelihoods: practical concepts for the 21st century. Institute of development studies (UK).

Ghabru MG, Devi G and Singh R. 2017. Agricultural Economics Research Review 30(1): 125-131

Girish CE, Kadian KS, Meena BS and Mandi K. 2020. An assessment of livelihood security of farmers practicing sericulture based dairy farming in Karnataka state. International Journal of Livestock Research 10: 43-50

Hatai LD and Sen C. 2008. An economic analysis of agricultural sustainability in Orissa. Agricultural Economics Research Review 21: 273-282

Hogger and Ruedi. 2004. In search of sustainable livelihood systems: Managing resources and change. Ruedi Baumgartnet and Ruedi Hogger, pp1-381, Sage publications, New Delhi

Makal A, Banerjee A, Roy A, Hazra S and Polley K. 2017. Issue and problems in agricultural development: A study on farmers of West Bengal. In Agricultural Anthropology P 179-189

Meena BS and Singh B. 2010. Impact of training programmes imparted by Krishi Vigyan Kendras in Rajasthan. International Journal of Agricultural Sciences 6: 213-215

Meena LK, Bairwa SL, Lakra K and Sirohiya L. 2015. Analysis of the profile on participating and non-participating farmers in chickpea production technology. Agriculture Update 9: 31-36

Noor KBM and Dola K. 2011. Investigating training impact on farmers' perception and performance. International Journal of Humanities and Social Science 1

Saha D. 2013. Role and scope of KVKs in development of aquaculture in rural areas. Conference paper

Shaheen A and Sanzidur R. 2012. Investigating livelihood security in poor settlements in Bangladesh. Annual conference of Agricultural Economics Society, University of Warwick, United Kingdom.

Sharma VK, Vaid A, Sharma PK, Ajrawat B, Jamwal A, Sharma N, Mahajan V and Gupta S. 2017. Impact assessment of training on farmer's perception, performance and entrepreneurship development. Maharashtra Journal of Agricultural Economics 20: 154-156

Singhal S and Vatta L. 2017. Impact of Krishi Vigyan Kendra on adoption of improved agricultural production practices. International Journal of Science, Environment and Technology 6: 993 – 1000

Singh DK and Singh P. 2014. Effectiveness of training programmes under Agricultural Technology Management Agency in Bihar. Indian Research Journal of Extension Education 14: 93-95

Singh K. 2000. Education, technology adoption and agricultural productivity. Indian Journal of Agricultural Economics 55: 473-489

Srinivas A and Shailja A. 2013. A review on impact of training programmes conducted by different training institutes. Indian Journal of Research 2: 266-267

Venu BN, Umesh KB and Gujanana TM. 2018. Livelihood security of agricultural labour households in rainfed region of north-Karnataka-An economic analysis. Indian Journal of Agricultural Research 52: 463-471

15

The Role of Krishi Vigyan Kendras (KVKs): A Critical Analysis

Jagannath Patra

Krishi Vigyan Kendra, Mayurbhanj-II, Jashipur, Odisha

Introduction

A large number of institutions in the field of agriculture and allied sectors are contributing to research in development of high yielding varieties of crops, technological innovations and other initiatives to boost production and human resource development. The technology available has to be permeated depending upon the necessities of the region – its soil, climate, culture and needs and means of the farmers, available human resources, feasibility and viability of different parts of the country. It has been observed that there are variations in knowledge and technological percolation. The National Commission of Farmers (founded in 2004) raised the issue of knowledge deficit, which directly impinges on agriculture productivity. There are significant gaps in backward and forward linkages between the agricultural laboratories and the farmers, in so far as transfer of technology is concerned. The 10th and 11th plans have emphasized the need for effective extension services. The 11th Plan Approach Paper also states that "in the longer run, growth in agriculture productivity can be sustained only through a continuous technological progress". The 11th Plan shows a concern towards the problem of transfer of technology and knowledge at grass-root level and puts forth challenges before the extension agencies. The 11th Plan points out that extension services are to be treated as a service delivery mechanism. The 12th Plan also emphasizes the need for mechanisation of agriculture with robust extension services.

The National Bureau of Soil Survey and Land Use Planning (founded in 2005) has estimated that about 146 million hectares of area is suffering from various types of land degradation which include water erosion, flooding, salinity, soil acidity, etc. About 62 per cent of the GDP in agricultural sector comes from crops (including horticultural crops). Livestock sub-sector contributes about

22 per cent, forestry about 10 per cent and fisheries about 5 per cent. Available data indicate a slight shift in the distribution away from the crop sector towards livestock and fisheries sub-sectors (Rao, 2014).

India is moving from primitive agriculture to modern mechanized farming but the change is very slow and only in some pockets of the country. Projections for subsequent years on the basis of relevant Annual Compound Growth Rate about 250 improved agricultural equipments and technologies have been designed and developed in India for various pre-and post-harvest operations operated by human, animal, mechanical and electrical power, for timely field operations (Pandey, 2006). Italia (2010) reported that farm mechanization (traditional to modern) takes place through activities at three levels, namely, village craftsmen, small industries and organized big industries. Despite all these efforts, change has not yet percolated to various sections of the society uniformly. Climate change is another area of concern for the agriculture sector. Vagaries of nature like floods, draughts, unusual and untimely rainfall in the country lead to low agriculture production and productivity. Indian Network of Climate Change Assessment (INCCA) study (Naresh Kumar et al., 2010) indicates that there could be a rise in the sea level, increase in cyclonic intensity, reduction in crop yield in case of the crops which are depending on rains, and reduction in milk. Inter-governmental Panel on Climate Change (IPCC)(2010) has predicted that productivity of most of the crops may decrease by 2 to 10% by 2020 and up to 30% by 2050 due to climate change. According to successive agriculture censuses, average operational landholding size (all categories) has declined from 2.82 hectare in 1970-71 to 1.16 hectare in 2010-11. Small and marginal holdings of less than 2 hectare account for 85% of the total operational landholding (Agriculture Census, 2010-11). The above discussions clearly indicate that there is a need for adequate transfer of technology and knowledge at grass root levels and also a continuous research in areas that are creating problems for the agriculture sector.

Extension, the Concept

Extension maybe defined as the science of developing capability of the people for sustainable improvement in their quality of life.

The use of the term `extension' originated in England in 1866 and taken up first by Cambridge and Oxford Universities. The term `extension education' was first used in 1873 by Cambridge University to take the educational advantages of universities to ordinary people. According to Ensminger (1957), extension is education and that its purpose is to change attitudes and practices of the people with whom the work is done. Leagans (1961) conceptualized extension education as an applied science consisting of content derived from research, accumulated field experiences and relevant principles drawn from the behavioural sciences synthesized with useful technology into a body of philosophy, principles, content and methods focused on the problems of out-of-school education for adults and youth.

Extension Strategy for Agricultural Development

The definite objectives of Agricultural Extension in the present era will be: Helping farmers to increase their standards of living by increasing farm income & yields of crops and animals by combining farm and non-farm sources of income. For Sustainable Rural livelihood and developing the farms, farmers need not only financial capital but also social, physical, natural, and a healthy human capital (Sustainable rural livelihood framework of DFID).

Agricultural extension has a crucial role to play in the context of growing demands of agricultural production in a sustainable manner. Reforms in the system envisage an extension service more broad-based and holistic in content and scope, thus beyond agricultural technology transfer. Its normal task of transferring and disseminating appropriate technologies and agronomic practices would not be sufficient. Extension agencies, services and functionaries will need to exercise a more proactive and participatory role, serve as knowledge information agents, initiating and facilitating mutually meaningful and equitable knowledge based transactions among primary producers, agricultural researchers and trainers. All this needs to be done in an effective and cost efficient manner.

The need for reforms in Agricultural Extension has been explicitly raised in the National Agriculture Policy; the report of Expenditure Reforms Commission, as well as, the Tenth Plan Approach paper. Keeping the recommendations of these policy initiatives in view, and to provide policy directives for extension reforms, a broad Policy Framework for Agricultural Extension (PFAE) has been developed by the Ministry of Agriculture, Govt. of India.

The five major guiding elements of the Policy Framework are as follows:

- Reforming Public Sector Extension
- Promoting private sector to effectively complement, supplement and wherever
- possible to substitute public extension
- Augmenting Media and Information Technology Support for Extension
- Mainstreaming Gender Concerns in Extension
- Capacity Building/ Skill up-gradation of farmers and extension functionaries

Evolution of KVK

Agricultural innovations and diffusion of new technologies are key drivers to attain food security in the country besides providing farmers a competitive edge over traditional farming, thus facilitating better standards of living. To realize their true potential, farmers must have access to the state-of-the-art technologies, necessary inputs and related information in all the segments, be it crop, livestock, forestry or fisheries. In this context, the Government of India through Indian Council for Agricultural Research (ICAR) has established a wide network of Krishi Vigyan Kendras (KVKs) in all the rural districts of the country. These KVKs under the aegis of the National Agricultural Research and Education System are the real carriers of frontline technologies and impart knowledge and critical input support for the farmers.

KVK as single window server for agriculture and rural development

The KVKs could provide this much required single window service to farming community as these centres are within the farmers' mobility capabilities. The SAU-KVKs are normally located at approachable distance to get technical advise from experts who are involved in research and to serve as a single window delivery system for the University in the rural areas.

Progress of Krishi Vigyan Kendra

The Education Commission (1964-66) recommended that a vigorous effort should be made to establish specialized institutions to provide vocational education in agriculture and allied fields at the pre-and post-matriculate levels to cater to the training needs of a large number of boys and girls coming from rural areas. The Commission, further, suggested that such institutions be named as 'Agricultural Polytechnics'. The recommendation of the Commission was thoroughly discussed: during 1966-72 by the Ministry of Education, Ministry of Agriculture, Planning Commission, Indian Council of Agricultural Research (ICAR) and other allied institutions. Finally, the ICAR mooted the idea of establishing **Krishi Vigyan Kendras (Agricultural Science Centres)** as innovative institutions for imparting vocational training to the practicing farmers, school dropouts and field level extension functionaries. The ICAR Standing Committee on Agricultural Education, in its meeting held in August 1973, observed that since the establishment of Krishi Vigyan Kendras (KVKs) was of national importance which would help in accelerating the agricultural production as well as in improving the socio-economic conditions of the farming community, the assistance of all related institutions should be taken in implementing this scheme. The ICAR, therefore, constituted a committee in 1973 headed by Dr Mohan Singh Mehta of Seva Mandir, Udaipur (Rajasthan), for working out a detailed plan for implementing this scheme. The Committee submitted its report in 1974. The first KVK, on a pilot basis, was established in 1974 at Puducherry (Pondicherry) under the administrative control of the Tamil Nadu Agricultural University, Coimbatore. The Planning Commission approved the proposal of the ICAR to establish 18 KVKs during the Fifth Five Year Plan. With the growing demand for more such Kendras, the Governing Body (GB) of the Council approved 12 more KVKs in 1979 and they were established in the same year from Agricultural Produce Cess Fund (AP Cess Fund). Owing to pending clearance of the Sixth Five-Year Plan scheme on KVK by the Planning Commission, the GB of the Council again approved 14 KVKs in 1981, which were established during 1982-83 from AP Cess Fund. A High Level Evaluation Committee on KVK was constituted by the ICAR in 1984. After thorough review of the programme, the committee strongly recommended for the establishment of more KVKs in the country. Keeping this in view the Planning Commission approved the establishment of 44 new KVKs during the Sixth Plan. Thus by the end of Sixth Plan, 89 KVKs had started functioning in the country. During the Seventh Plan, 20 more KVKs were established. The success of the KVKs at many locations created a great demand for establishment of new KVKs in the remaining districts of the country. Accordingly, the Planning Commission further approved 74 new KVKs to be established during the period 1992-93. Again in the Eighth Plan (1992-97), 78 new KVKs were approved and established in the country, taking the total number of functional KVKs by the end of the Eighth Plan to 261. The number of KVKs increased to 290 during Ninth Plan with the establishment of 29 more KVKs. On the occasion of the Independence Day Speech on 15th August 2005 the Prime Minister of India announced that by the end of 2007 there should be one KVK in each of the rural districts of the country. This resulted in 551 KVKs by the end of the Tenth Plan. At present there are 700 KVKs.

Krishi Vigyan Kendra (KVK)-Farm Science Centre

Vision

Science and technology-led growth leading to enhanced productivity, profitability and sustainability of agriculture

Mission

Farmer-centric growth in agriculture and allied sectors through application of appropriate technologies in specific agro-ecosystem perspective

Mandate

The mandate of KVK is **Technology Assessment and Demonstration** for its **Application** and **Capacity Development.**

a) On-farm testing to assess the location specificity of agricultural technologies under various farming systems.

b) Frontline demonstrations to establish production potential of technologies on the farmers' fields.

c) Capacity development of farmers and extension personnel to update their knowledge and skills on modern agricultural technologies.

d) To work as Knowledge and Resource Centre of agricultural technologies for supporting initiatives of public, private and voluntary sectors in improving the agricultural economy of the district.

e) Provide farm advisories using ICT and other media means on varied subjects of interest to farmers.

In addition, KVK would produce quality technological products (seed, planting material, bio-agents, livestock) and make it available to farmers, organize frontline extension activities, identify and document selected farm innovations and converge with ongoing schemes and programmes within the mandate of KVK.

To operationalise the mandated activities, following broad objectives of KVKs are

- To promptly demonstrate the latest agricultural technologies to the farmers as well as extension workers of State Departments of Agriculture/ Horticulture/ Fishery/ Animal Science/ NGOs with a view to reducing the time lag between the technology generation and its adoption.
- To test and verify the technologies in the socio-economic conditions of the farmers keeping in view the production constraints and to modify the technologies to make them appropriate
- To impart training to the practicing farmers/farm women, rural youth and field level extension functionaries by following the methods of "Teaching by doing" and "Learning by doing"
- To back-up with training and communication supports to the district level development departments viz. Agriculture/Horticulture/Fisheries/Animal science and NGOs in their extension programmes

New Initiatives in Krishi Vigyan Kendras

Interface

- Organization of Kharif and Rabi Kisan Sammelans
- Organization of Jai Kisan – Jai Vigyan Week
- Mera Gaon Mera Gaurav Initiative: 10712 villages regularly visited by scientists of ICAR Institutes/SAUs to provide advisories to the farmers

ICT Initiatives

- Web Portal- Krishi Vigyan Kendra Knowledge Network Portal, dynamic portal having read and write facility, launched on 8 July 2016 for regular monitoring of KVKs and to provide information and advisories to the farmers
- mKisan Portal- 90 lakh farmers are being provided advisory by the KVKs

Strengthening KVKs

- Project on "Promotion of Agricultural Mechanization for *In-situ* Management of Crop Residue" has been initiated during 2018-19 involving 60 KVKs with focus on promotion of agricultural machineries, demonstrations, training and other Information, Education and Communication (IEC) activities.
- Focused programmes under Knowledge Systems and Homestead Agriculture Management in Tribal Areas (KSHAMTA) in 125 districts; and Nutri-sensitive Agricultural Resources and Innovations (NARI) as well as Value Addition & Technology Incubation Centres in Agriculture (VATICA) in 100 districts have been initiated by KVKs.
- Since 2016-17, the KVKs are organizing National Skill Qualification Framework aligned skill development training programmes related to agricultural and allied sectors with financial support from Department of Agriculture, Cooperation & Farmers' Welfare.
- MoU has been signed with IMD and Bureau of Energy Efficiency for backstopping to KVKs on climate change aspects.
- Automatic Weather Station (AWS): AWS established at 121 KVKs under National Innovations in Climate Resilient Agriculture (NICRA) Project and 100 being established with support from India Meteorological Department
- Establishment of 121 Climate Smart Villages under National Innovations in Climate Resilient Agriculture
- Integrated Farming System models at 100 KVKs for capacity development of farmers

- Establishment of Pulses Seed Hubs in 97 KVKs
- Attracting and Retaining Youth in Agriculture (ARYA): 100 KVKs are implementing this project to attract and empower the youth in rural areas for taking up various agriculture and allied sector enterprises.

Participatory Technology Development

- Farmer FIRST Programme-51 projects being implemented by ICAR Institutes and SAUs/CAU for enhancing Farmers–Scientist interface

Review of Literatures

Agricultural extension services in the country have reported to be demand driven, participatory and decentralized in which accountability is geared toward the users (Birner et al., 2006; Biurner and Anderson, 2007; Davis 2008; Hall et al., 2000; Kokate et al., 2009; Sulaiman and Hall, 2008; Swanson, 2009). Country"s five year planning emphasized the role of agricultural extension services in increasing agricultural growth. The five year plans emphasized the need for strengthening the agricultural extension in the country.

The NSSO survey (2003) indicated that 60% farmers had no access to any source of information on new farming technology. Among the farmers who had accessibility, 16% got it from progressive farmers and input dealers (NSSO, 2005). The relevance of the information provided to the extension services was also in question. There are multiple sources of information flow to the farmers including progressive farmers, input dealers, print and electronic media and extension workers. The study by Adhiguru, Birthal, and Ganesh Kumar (2009) indicates that small farmers had to depend upon progressive farmers, input dealers and radio for information. Contact with extension workers for medium size and large scale farmers was almost double that of small holding farmers. Sulaiman"s*et al* work on agricultural extension suggests that extension services should be strengthened. The work examines the processes and structures inherent in the organizations that provide extension (Sulaiman and Holt, 2002; Sulaiman and Hall, 2002; Sulaiman, 2003a; Sulaiman and van den Ban, 2003; Sulaiman, 2003b; Sulaiman and Hall, 2008). A review of agricultural extension in India by Claire *et al.* (2010) brings out that extension services in India need to evolve to provide a diverse set of services and outreach to marginal and small farmers. Extension services should respond to emerging issues in agriculture. The research by Venkatesh and Nithyashree (2014) examines the inputs used in agriculture and its accessibility. The results indicate that while input use had expanded in the second half of the 2000s, the role of private sector was more visible in supplying inputs like seeds, fertilizers and pesticides etc. The findings also suggest that inclusion of small and marginal farmers under institutional credit coverage and special attention of extension system to reach the unreached farmers are necessary. Impact study of KVKs on beneficiaries of Tamil Nadu and Puducherry by Subburaj (2013) covers six KVKs and 3,000 beneficiaries. The study reveals that while services provided by KVKs were useful to the beneficiaries, there was a need for frequent visits by the KVK staff to the field level. About 16.8% beneficiaries stated that KVK staff needs up-gradation of their technical knowledge for providing guidance on various problems faced by the farmers.

Analysis Supplemented with Case Illustrations

To find out the performance of the KVKs, a study was done on the KVKs of the Odisha. From the record, it is stated as there are 33nos of KVKs in Odisha has marked their presence in 30 districts of the state in which there are additional KVKs in three districts i.e. Mayurbhanj, Ganjam and Sundargarh. Out of which 31 are operating under Odisha University of Agriculture and technology whereas two are under NRRI and CIFA directly under the administrative control of ICAR.

The study was conducted by content analysis from various reports obtained from the individual websites alongwith a visit to one KVK i.e. KVK, Mayurbhanj-1 at Shmakhunta for accessing the reality at ground level by personal interaction with scientists, farmers and other stake holders.

The Krishi Vigyan Kendra came into existence at Shamakhunta in Mayurbhanj District on 28th MAY 2005 as per the sanction of ICAR vide letter No. 3-56-93-AE, dt. 19.05.2005. The organisation caters to the need of the Mayurbhanj District located in the North central plateau Agro-climatic Zone of the State. The Kendra is functioning under the administrative control of the Orissa University of Agriculture & Technology, Bhubaneswar and fully financed by the Indian Council of Agricultural Research, New Delhi. The operation area of KVK covers 14 no of blocks of the district.

Thrust Areas

Kharif Paddy

- Varietal substitution in medium and low land paddy
- Crop diversification in upland paddy
- Seed production programme
- Weed management
- Integrated Pest & Disease management

Oilseeds and Pulses (Ground nut (Rabi), Green gram (Rabi) and Arhar)

- Varietal replacement with recommended fertilizer and need based plant protection Techniques
- Liming and Bio-fertilizer application
- Weed management

Vegetables

- Off season vegetable cultivation
- Varietal substitution and IPM
- Organic vegetable cultivation
- Storage of vegetables

Fruits

- Development of mango orchard and nursery
- Tissue culture banana cultivation
- Quality planting material production
- Post harvest technology including grading and packaging

Dairy, Poultry and Goatery

- Backyard improved poultry farming
- Deworming and supplementary feeding in Goatery
- Dairy farming and fodder cultivation

Fishery

- Stocking of advanced fingerling in community pond management
- Multiple stocking and harvesting
- Yearling production

Other Enterprises

- Mushroom Cultivation
- Commercial floriculture
- Bee-keeping and lac cultivation
- Vermicomposting
- Value addition of fruits and vegetables
- Farm mechanization

Facilities Available

Vermicompost Unit, Nursery, Mushroom Unit, Mushroom Spawn production Unit, NADEF Compost Unit, Crop Cafeteria, Poly house and Shade net, Demo units and medicinal Garden, Mango Orchard, IFS unit.

Highlight on the Broad Dissection of KVKs in Odisha

- Almost all the selected KVKs reported that agricultural crops, horticulture and home sciences were the focus areas of their activities. Few of the KVKs also referred to veterinary, fisheries sciences and forestry as their focus.
- In general the KVKs are short of staff at all levels among Scientists and Technical Support staff.
- KVKs have reported that they are partly satisfied regarding sanctioned infrastructure facilities. However, in a number of KVKs infrastructure is not available as per norms especially staff quarters, hostels, equipments etc.
- Off-campus activities are more than on-campus activities. The data also show that KVKs are concentrating more on individuals than the groups.
- KVKs are responding well to requests from farmers for information-attending to 90 per cent of the requests received in a year. The KVK also responds to requests for seeds and planting material, quality animals and assistance in implementing technologies reasonably..
- Most of the KVKs felt that rich farmers, those located near the KVK, those with secondary or higher education and those whose land holdings were comparatively large were more likely to implement new technologies offered by the KVKs. While the linkage between richness/poverty and proneness to adopt technologies and that between size of holding and proneness to adopt is strong, the effect of education on proneness is somewhat weak. It is perhaps the resources to adopt that matter more than education.
- On an average about 40% time each is devoted to non mandated activities
- KVKs reported an edge over other organizations providing technology services in terms of having demonstration units, a team of technical experts in varying disciplines, advice to the farmers under one umbrella, use of ICT and provide mobile advisory services.
- Most of the KVKs were of the opinion that a number of technologies were gender sensitive and had helped in reduction of drudgery, income enhancement and developing self-confidence among women thus making them empowered.
- KVKs have limited outreach and catering more to the needs of big and progressive farmers KVKs are the front runners among other organizations with half of the farmers acknowledging that training and demonstrations organized by those institutions accounted for the technology transfer. For making KVKs more effective farmers indicated that the present activities and mandate needs to be reviewed and expanded.

It has been observed that where resources are scarce or farmers are unable or unwilling to participate in various activities, the results of outputs are difficult to achieve. KVKs have expertise so far technical knowledge in various agro-related fields is concerned, officials are capable of providing training to farmers, infrastructural support ranges from inadequate to adequate and inputs in terms of new and emerging technologies are available with KVKs in general. These inputs helped in organizing various activities by KVKs which resulted in direct outputs in terms of OFT, and FLD, farm development in the premises of KVKs for display and training, technology transfer and advice to farmers.

The collaboration with other departments by KVKs in general has also been observed. Up to this level it can be concluded that KVKs are able to achieve the outputs to a great extent according to their objectives and mandates. When it comes from outputs to outcomes, the study indicates that KVKs are playing an important role in achieving the outcomes. Women have been empowered and the spill over effect of the new technology has also been observed as the success of new technology disseminated to some farmers is being seen by other fellow farmers and they are also adopting the same. This indicates that the assumption that farmers will be willing to use new technology and new knowledge and will have accessibility to the new technology has been validated to certain extent. An attempt was made to study the long term impacts of interventions made by KVKs. The results show that in some cases the quality of life has improved as the enhanced income is being used for better education of children, accessing better health services and spending money in assets creation, etc. While in some other cases such type of results could not be achieved as enhanced incomes were used appropriately.

The KVK services are limited to training programmes on farm trial and field demonstration and do not come as a complete package while farmers require the services in a complete package that include forward and backward linkages. The results of the survey and interaction indicate that technology adoption has a limited success if the new technology and inputs are not available or accessible and also if the post-harvest technology assistance in marketing, etc. is not available. other studies conducted on these issues (IFPRI, 2010) indicate the same pattern of results. There is an immediate need to review the mandated activities of KVKs.

The role of mass media and information and communication technology (ICT) has been found very useful in disseminating the new technology in agriculture. It has been found that the KVKs that are collaborating with mass media or are using ICT have been successful in transferring of technology with a wider coverage of farmers. Such efforts should be incentivized and promoted if the effectiveness and efficiency of KVKs has to be enhanced.

The results of the study indicate that there is a need to have intervention not on individual basis (farmer level) but at group/ family level where the needs of the families are taken into focus, for example, the adult males involved in farming should be provided with latest technology while youth and women folk should be provided entrepreneurial training. In this way, on one hand women will be empowered socially and economically, youth will be able to establish value chain by establishing micro enterprises using agriculture produce.

The results show that technology transfer had been effective and efficient wherever there are farmers" clubs, farmers" associations and other farmers" groups continuously interacting with KVKs and other extension personnel. The results of the present study also reveal that the learning from fellow farmers had been very effective.

Various studies have brought out the efficacy of extension services in the country. The situations in agriculture sector are fast changing. The technology in agriculture is evolving. Is this new knowledge being implemented at grassroot level and are farmers accepting the same? If yes, then what is the impact in terms of increased productivity in agriculture, increased incomes of farmers resulting in better quality of life? Is the development sustainable? How much time it takes to transfer the technology? What else is needed to make KVKs more effective? These are some of the questions that need to be addressed. It is also to be seen that, if new practices are not being adopted by the farmers then what are the reasons for that?

Summary and Conclusion Including Suggestions

Infrastructure and Resources

- Adequate infrastructure and resources, both in human and monetary terms, provide the essential base for carrying out various activities. There is at present a uniform pattern of distribution of staff and resources across KVKs. The study suggests that resource distribution should be as per the specific needs of the KVK taking into account the geography of the area, location of KVK, local problems and prospects.
- More frequent field visits are needed during the agricultural season which requires not only human resources but also financial resources. KVKs should be strengthened by adding more field staff exclusively for extension visits
- Over the years there is a reduction in budgetary allocations which is leading to reduction in the coverage of the KVKs" activities. To make KVKs effective, adequate budget should be allocated. There are also delays in sanctioning of budgets which leads to financial crunch hampering the activities.

- Training is one of the mandated activities of KVKs but the budget sanctioned for this purpose is not adequate; per day allocation/allowances for farmers/ trainees are too meager. Budget needs to be enhanced if training is to be effective.
- Majority of the farmers who were not benefited by KVK programmes feel that the inconvenient location of the KVKs is one of the main reasons for not accessing their services.
- Inadequate and erratic power supply is a big problem in a number of KVKs and is affecting their efficiency. There needs to be a provision for alternative power supply through generators.
- KVKs that are in remote places and hilly and difficult regions need special consideration. Some incentives should be given to KVK staff in such cases.

Policy Issues

There is an urgent need of capacity building of KVK staff. A training policy should be adopted and staff provided need-based training. It was observed during the study that some of the farmers had more up- dated knowledge than KVK personnel.

- The career advancement promotion and transfer policy for the staff of KVK needs a re-look. There is no career advancement for the scientific staff of KVK and that is one reason why the scientists do not want to join this organization while some others are leaving it. It was noticed at many places that there were frequent transfers of the personnel that affects the efficient functioning of KVKs.
- To give more field assistance to farmers there is a need for more middle and lower level extension workers with practical knowledge for guidance and regular contact and follow ups and also cater to the needs of small and marginal farmers on regular basis.

References

Adhiguru P., Birthal P.S., and B. Ganesh Kumar (2009), Strengthening Pluralistic Agricultural Information Delivery Systems in India. Agricultural Economic Research Review

Aggarwal Abha and Pandey Arvind (2006): Lot Quality Assurance Sampling in Evaluation of Immunization Programme published in Biostatistical Aspects of Health and Population ISBN 81-7075-081-4

Agrawal, Rashmi, (2015). "An Overview of Methodologies in Evaluation, paper presented in a workshop on Evaluation for Development of Equity and Equality" organized by Institute of Social Studies Trust, and NITI Aayog, New Delhi, April 16-17.

Agrawal, Rashmi, D. Rama Rao and Sachita Bhattacharya (2012). "Planning Human Resources for Inclusive Development of Small and Marginal Farmers", The Indian Economic Journal, Vol. 60(3), October-December 2012.

Agrawal, Rashmi., Joshi G.P., Indrakumar and Bhattacharya, Sanchita. (2013). "Human Resource Planning for Agricultural Growth" in Agriculture Education & Knowledge Management, edited by B.S. Hansra, P.K. Jain, Suresh C. Babu and V.K. Bharti. D.P.S. Publishing House, New Delhi.

Agri. Census (2011),Agriculture Census Division Department of Agriculture & Co-Operation Ministry of Agriculture Government of India

Birner, R., and J.R. Anderson. (2007). How to make agricultural extension demand-driven? The case of India"s agricultural policy. IFPRI Discussion Paper 00729. Washington, D.C.: International Food Policy Research Institute.

Chapman, R., and Slaymaker, T. (2002). ICT and rural development: Review of the literature, current interventions and opportunities for action. Working Paper 192. London: Overseas Development Institute.

Claire J. Glendenning (2010). Review of Agricultural Extension in India Are Farmers" Information Needs Being Met? IFPRI Discussion Paper 01048.

Clark, N., J. Smith, and Hirvonen, M. (2007). Livestock R&D in east and southern Africa: An innovation systems perspective with special reference to the International Livestock Research Institute. International Journal of Technology Management and Sustainable Development. 6(1): 9-23.

Collion, M-H., and Rondot, P. (1998). Partnerships between agricultural services institutions and producers associations: myth or reality? Agricultural Research and Extension, Network Paper No.80. London: ODI.

Gandhi, R. (2009). Digital Green: Participatory video and mediated instruction for agricultural extension USC Annenberg School for Communication, 5(1): 1-15.

GoI (2015). Union Budget of India, 2015-16

GoI. (2001). Tenth Five Year Plan Approach Paper, Planning Commission, Government of India, New Delhi

GoI. (2006). Eleventh Five Year Plan Approach Paper: Towards Faster and More Inclusive Growth, Planning Commission, Government of India, New Delhi

IFPRI, (2010). "Review of Agricultural Extension in India – Are Farmers" Information Needs Being Met? International Food Policy Research Institute Discussion Paper 01048, Eastern and Suthern Africa Regional Office. Italia, 2010.

Kumar, N. (2010). Indian Network of Climate Change Assessment (INCCA)

National Bureau of Soil Survey and Land Use Planning (2005) Impact Evaluation: Methodological and Operational Issues, Asian Development Bank, 2006

NSSO (National Sample Survey Organisation) (2005). Situation assessment survey of farmers: Access to modern technology for farming, 59th round (January-December 2003), Report No. 499 (59/33/2). New Delhi: Ministry of Statistics and Programme Implementation

Ramarao D, Agrawal Rashmi, Nanda S.K., Awasthi I.C., Joshi G.P., Bhattacharya Sanchita, Indra Kumar (2011). "Human Capital Requirements in Agriculture and Allied Sectors, NAIP Project Report, National Academy of Agricultural Research Management, Hyderabad, India published by Institute of Applied Manpower Research, New Delhi.

Rao BVLN. (2014). Indian Agriculture Scenario in Towards Skill Endowed Human Resources in Agriculture (2014) by D. Rama Roy, BNLN Roy, Rashmi Agrawal (eds.).

Subburaj V.K. (2013). Impact of Krishi Vigyan Kendra"s on the Beneficiaries in Tamil Nadu and Puducherry, Ph.d. thesis, Gandhigram Rural Institute, Gandhigram (Tamil Nadu).

Sulaiman, R. (2003). Innovations in Agricultural Extension in India. Rome: Food and Agriculture Organization of the United Nations, Sustainable Development Department. 112.

Sulaiman, R., and A. Hall. (2002). Beyond technology dissemination: Can Indian agricultural extension re-invent itself? ICAR Policy Brief 16. New Delhi: Indian Council of Agricultural Research

Sulaiman, R., and A.W. Van den Ban. (2003). Funding and delivering agricultural extension in India. Journal of international Agricultural and Extension Education. 10(1): 21-30.

Thapa, G. and Gaiha, R. (2011). Smallholder Farming in Asia and the Pacific: Challenges and Opportunities, Paper presented at the Conference on New Directions for Small Holder Agriculture, January 24-25, Rome: IFAD.

Venkatesh P and Nithyashree M.L. (2014). Institutional Changes in Delivery of Agricultural Inputs and Services to Farm Households in India, Agricultural Economics Research Revieq. 27(Conference).

White Howard (2009). International Initiative For Impact Evaluation, Theory-Based Impact Evaluation: Principles and Practice, 3ie Working Paper 3, New Delhi.

16

A Study on Constraints Perceived by the NGO (Non Government Organization) Led SHGs (Self Help Groups) Members in Sonepur District of Odisha

Subarnna Keshari Haripriya Padhan[1] and Debasis Jayapuria[2]

[1]Bidhan Chandra Krishi Viswavidyalaya, Mohanpur, Nadia, West Bengal
[2](Ag. Extension), Krishi Vigyan Kendra, Mayurbhanj - I, Odisha

Abstract

Women empowerment is the most important instrument for the socio-economic development of a nation. In this context, SHGs have emerged as the tools that wield power to create a socio-economic revolution in the rural areas of our country. The empowerment of rural women is a key priority in the National Development Policy of India. The present study was taken in Sonepur district which has selected purposively for the study on the basis of maximum number of SHGs. The study was conducted on constraints perceived by the NGO led SHGs members. The district consists of six blocks, out of which two blocks namely Ullunda and Birmaharajpur were selected purposively for the study. A proportionate random sampling procedure was used for selection of the SHGs. A list of SHGs were obtained from records. From the prepared list, random sample was selected from two blocks. Therefore, 15 SHGs from each block were selected randomly. Thus, the total SHGs size was 30 SHGs. From each selected SHGs, two women members (president and secretary) were interviewed. Thus, the total members of women respondents from the 30 SHGs were 60. Thus the total number of respondents for the present study was 60. Data were collected with the help of interview schedule by the researcher and for analysis of data frequency, Mean Percent Score and rank were adopted. The results was shown that major constraints expressed by green gram growers under knowledge and information constraints was lack of information on post harvest technology where as in socio economic constraints, low profit and in technological constraints, lack of market linkage were ranked among major constraints.

Keywords: *Constraints, Mean percent score, women, Empowerment, SHG (Self Help Group), Adoption and Rank.*

Introduction

Women play a pivotal role in agricultural and rural economies in all developing countries. Rural India woman is facing Poverty and unemployment is the major problems. In India at the end of ninth five year span 26.1% of the population was living below poverty line. In the rural area 27.1% of the population was living under poverty. The overall unemployment rate is estimated to 7.32%. The female unemployment rate is 8.5%. The rate of growth of women unemployment in the rural area is 9.8%. A series of development programmes have been implemented for the development of economy in general ignoring women who constitute 50 percent of the total population. Role of women in development is an indispensable factor. Development programmes no longer can achieve their target without the participation and contribution of women. Women empowerment is multifaceted concept that extends to the psychological, economic, social, cultural, political and institutional spheres of women's lives. The Government of India has launched and implemented a number of schemes towards poverty alleviation and women empowerment but it was observed that women in rural areas especially from the poor families could not be benefitted. Women empowerment is the most important instrument for the socio-economic development of a nation. In this context, SHGs have emerged as the tools that wield power to create a socio-economic revolution in the rural areas of our country. The empowerment of rural women is a key priority in the National Development Policy of India. In India, the organization of SHGs, especially those promoted by Non Governmental organizations (NGOs) is widely recognized as a development strategy for poverty reduction. NGOs also known as voluntary sector have become an irresistible global force today and it is growing in relation to its presence in developmental activities. Basically, an NGO or voluntary organisations are non-profit making agencies that are constituted with a vision by a group of likeminded people, committed for the uplift of the poor, marginalized, unprivileged, underprivileged, impoverished, downtrodden and the needy and they are closer and accessible to the target groups, flexible in administration, quicker in decision making, timely in action and facilitating the people towards self reliance ensuring their fullest participation in the whole process of development. SHGs are started by NGOs that generally have broad anti-poverty agendas. SHGs are seen as instruments for goals including empowering women, developing leadership abilities among poor people, increasing school enrollments and improving nutrition and the use of birth control. SHGs are small group of individual members who voluntarily come together and form an association for achieving a common objective. In most cases, SHGs are constituted by persons known to one another and coming from the same village community or neighbourhood. SHGs are small in size with membership ranging from 10 to 20 and are homogeneous. A strengthening development among women the nation over has been currently turned by quick advancement in SHG arrangement. The arrangement of SHGs isn't conclusively a miniaturized scale credit venture however a strengthening procedure. The strengthening of women through SHGs would give advantage not exclusively to the individual women yet in addition for the family and network all in all through aggregate activity for improvement. Women constitute half of any country's human endowment. All round development and harmonious growth of a nation would be possible only when women are considered as equal partners in progress with men.

Methodology

Odisha state is divided into 30 districts. Out of 30 districts, Sonepur district was selected purposively for the study on the basis of maximum number of SHGs. Sonepur district consists of six blocks namely, Binika, Birmaharajpur, Dunguripali, Sonepur, Tarava and Ullunda, out of which two blocks namely Ullunda and Birmaharajpur were selected purposively for the study. A proportionate random sampling procedure has been used for selection of the SHGs. A list of SHGs was obtained from records. From the prepared list, random sample was selected from two blocks. Therefore, 15 SHGs from each block were selected randomly. Thus, the total SHGs size was 30 SHGs. From each selected SHGs, two women members (president and secretary) were interviewed. Thus, the total members of women respondents from the 30 SHGs were 60. Thus the total number of respondents for the present study is 60. Data were collected by the researcher by personal interview method with the help of interview schedule in the month of November and December 2019. In the present study it was intended to access the constraints perceived by the NGO led SHGs members.These constraints were categorized into four categories namely knowledge and information constraints, socioeconomic constraints and technological constraints. The intensity of constraints was measured on a three point continuum scale. Weight of 3 2 and 1 were given for most important, important and least important constraint, respectively. Total score obtained by each respondent as well as for each statement was calculated. Finally mean percent score was calculated by the following formula. Mean percent score (MPS) = Total score obtained/Maximum obtainable score × 100.

Results and Discussion

Knowledge and Information Constraints

It is evident from table 1 that three constraints were recorded such as lack of information on post harvest technology, lack of knowledge about value addition in fruits and vegetables and lack of training on post harvest technology and value addition of fruits and vegetables. Lack of information on post harvest technology has been ranked first among this group by the farmers (MPS=92.23). A key to the diffusion of a new technology is providing information about its existence and properties. The lack of technical knowledge as a constraint has given a scope to train farmers about the new package of practices for better results. More over the second constraint and third constraints are lack of knowledge about value addition in fruits and vegetables (MPS=89.45) followed by lack of training on value addition and post harvest technology of fruits and vegetables (MPS=83.89).

Table 1: Knowledge and information constraints faced by NGO led SHGs Members

(N=60)

Sl. No.	Constraints	Frequency			MPS	Rank
		Most Important (3)	Important (2)	Least Important (1)		
1	Lack of information on post harvest technology	51	4	5	92.23	I
2	Lack of knowledge about value addition in fruits and vegetables	47	7	6	89.45	II
3	Lack of training on value addition and post harvest technology of fruits and vegetables	41	9	10	83.89	III

Socio economic constraints

Table 2 reveals that four constraints were perceived by the NGO led SHGs member as related to their socioeconomic conditions. Low profit and lack of management were the two major constraints in this category. These were ranked I & II with an overall MPS of 92.22 and 90.00 respectively. The other constraints in this category were lack of women participation (MPS=83.33) and high cost of inputs (MPS=73.89).

Table 2: Socio economic constraints faced by NGO led SHGs Members

(N=60)

Sl. no.	Constraints	Frequency			MPS	Rank
		Most important (3)	Important (2)	Least important (1)		
1	High cost of inputs	31	11	18	73.89	IV
2	Lack of management	47	8	5	90.00	II
3	Lack of women participation	40	10	10	83.33	III
4	Low profit	50	6	4	92.22	I

Technological constraints

Table 3 reveals technological constraints were perceived by the SHGs members are Lack of market linkage and lack of post harvest machineries were the two major constraints ranking I and II of MPS 94.44 and 92.22 respectively. The other constraints in this category were poor bank linkage (MPS=90.56) and lack of timely technical advised (MPS=82.22) ranking III and IV respectively.

Table 3: Technological constraints faced by NGO led SHGs Member

(N=60)

Sl. no.	Constraints	Frequency			MPS	Rank
		Most important (3)	Important (2)	Least important (1)		
1	Poor bank linkage	45	13	2	90.56	III
2	Lack of post harvest machineries	48	10	2	92.22	II
3	Lack of timely technical advised	38	12	10	82.22	IV
4	Lack of market linkage	51	8	1	94.44	I

Conclusion

It can be concluded that major constraints expressed by NGO led SHGs members under knowledge and information constraints is lack of information on post harvest technology where as in socio economic constraints, low profit and in technological constraints, lack of market linkage ranked among major constraints. The constraints expressed should be taken care by the researchers, state agricultural departments personal, extension agencies and commercial firms to orient their infrastructure for higher adoption of better and scientific management for running of SHGs for maximum production. Proper guidance and awareness for the farmers should be created through practical skill oriented training, field visits, method demonstrations and through various extension literatures (printed as well as video). The maximum participation of women in technology generation process as well as in the production enhancements strategies are required to face any risk of technological failures.

References

Amutha, D. (2011). Socio-Economic Impact through Self Help Groups. Journal of Economics and Sustainable Development, 2(6), 89-95.

Banerjee, C. G. T. (2012). What Factors Play a Role in Empowering Women? A Study of SHG Members from India. Gender technology and development, 16(3), 329-355.

Dhruba Hazrika, "Women strengthening in India: A Brief Discussion", 2011, International Journal of Educational Planning and Administration, Vol.1, pp. 199-202.

Reji, (2013), "Economic Empowerment Of Women Through Self Help Groups In Kerala", International Journal Marketing, Financial Services & Management Research, 2(3): 97-113.

Femina, P. (2003), 'Role of voluntary organizations in empowering women through self help groups', unpublished MSc (Ag) thesis, TNAU, Coimbatore.

Ghosh, D.K. (2001). NGOs Intervention in Poverty Alleviation, Kurukshetra, March, pp. 2-9.

Jain, D., & Nai, P. (2013). SHG Helping Empower Rural Women – A Study. International Journal of Social Science Interdisciplinary Research, 2(7): 126-135.

Kotishwar, A. & Khan, M. A. A. (2010). "Women Empowerment through Self Help Groups – A Case Study of Nizamabad District Of Andhra Pradesh", International Journal of Research in Commerce and Management, 1(1): 48-71.

Kour. A.(2008). Self-Help Group (SHG) and Rural development, Kurukshetra, 57(2): 25-29.

Krishnaveni, V., Haridas, R., Nandhini, M., & Usha, M. (2013). Savings And Lending Pattern Of Help Groups: An Overview. i-Manager's Journal on Management, 8(1): 49.

Kumar, S.S. (2011). Women empowerment and poverty reductions through self help groups in the Nilgiri district. Golden Research Thoughts, I (1): 1-4.

Narayana Swamy, N.S Manivel and Bhaskar (2005). Micro credit driven rural enterprises lessons from SHGs Assisted under SGSY scheme. Journal of Rural Development , 24(3):353-376.

Ramachandran, T. and Balakrishnan, S. (2008). Impact of Self-Help Group on Women Empowerment – A study in Kanya Kumari district. Kurukshetra, 57(2): 31-34.

Savita, B. and Jyothi, P. (2012). Self Help Groups and their Role in Socio Economic Empowerment- A Study in the States of Andhra Pradesh and Tamil Nadu. Research Journal of Social Science and Management, 2(4): 78-88.

Sk. Asha B. (2014). Effect of Self-Helps on Empowerment of Rural Women-A Case Study in Guntur District Of Andhra Pradesh. GJRA, 3(8): 1-2.

Venkatesh, J. and Kala, K. (2010). Empowering rural women all the way through self-help groups. International Journal of Management, 1(2): 156-63.

17

Solar Energy Awareness and Adoption for Agricultural Operations: A Path Toward Green Energy Use

Aditya, Arulmanikandan B, Sube Singh and P.S. Shehrawat

Department of Extension Education, CCS HAU, Hisar, Haryana

Abstract

According to the most recent UN estimates, the population of the globe could increase to over 9.7 billion in 2050, and to fulfill the demands of an expanding population and a rising need for food, the energy demand in the agricultural sector has dramatically expanded. Additionally, other global issues like climate change, water shortage, and energy security require more environmentally friendly and sustainable farming methods. Using renewable energy sources, particularly solar energy, which is the most accessible and appropriate for agricultural use, is one way to meet the energy needs of agricultural operations. A survey was undertaken to study the level of awareness and adoption of solar energy for agricultural operations in the Hisar district of Haryana state. From the district, eight villages, and from each village 15 farmers were selected randomly. The study revealed that more than half (68.33%) of respondents were aware about 'solar energy is an eco-friendly green energy source' followed by 'no harmful effect of solar energy on humans and animals (67.50%)'. The adoption level of farmers about the use of solar energy in agricultural operations is very low with an overall adoption level of 18.36 per cent. It also discusses the limits associated with the adoption of solar energy. As a result, this study offers a pertinent perspective of previous farmers who have employed solar energy in their fields in order to comprehend the diverse perceptions that farmers have while adopting such innovation.

Keywords: Awareness, Adoption, Solar energy, Climate change, Energy security.

Introduction

Humans primarily get their nourishment from agriculture. According to the most recent UN estimates, the population of the globe could increase to over

9.7 billion in 2050, and to fulfill the demands of an expanding population and a rising need for food, the energy demand in the agricultural sector has dramatically expanded. However, a farm that produces food needs a steady supply of energy to run the machinery, irrigation pumps, and other devices that are often powered by fossil fuels. Because of the ongoing demand for energy, there is growing concern that these fossil fuels will also run out soon (Aroonsrimorakot*et al.,* 2020). Additionally, other global issues like climate change, water shortage, and energy security are just a few variables affecting agricultural production that require more environmentally friendly and sustainable farming methods As a result, it is crucial to maximize land usage and increase agricultural production through the deployment of cutting-edge farming techniques and new technology for both economic effectiveness and the best possible use of the limited resources available in the form of land. There are many distinct uses for renewable energy in agricultural systems. Renewable energy sources can also be used to generate electricity for a range of agricultural tasks (Torshizi and Mighani, 2017). Renewable energy can come from a variety of sources, but solar energy is the ideal option because it can be installed practically anywhere on the field, is pollution-free, and is also the most economical. Unlike the usage of fossil fuels, using solar energy doesn't release greenhouse gases. For remote locations with erratic access to electricity, it is the best renewable energy choice. Applications for space and water heating, greenhouse heating, solar photovoltaic systems, and water pumps for irrigation using various forms of solar energy technology (Mekhilef*et al.,* 2013). There are numerous benefits of using solar energy technologies, such as solar panel water pumping systems, aside from being free of fuel costs, solar panels also have lower maintenance costs and require fewer replacement components than generators powered by diesel engines. In order to meet the growing demand for agricultural products from the expanding population, agriculture is a sector that aspires to accept the usage of solar energy and reaps enormous benefits from it. Therefore, solar energy technology has emerged as a solution in order to boost economic benefits to farmers and improve the environment by reducing CO2 emissions into the atmosphere, thereby addressing the issues caused by a growing population and a shrinking amount of available resources (Aroonsrimorakot and Laiphrakpam, 2019). The potential for farmers to move from using fossil fuels to using renewable alternative energy, which might lead to produce superior in both quantity and quality of food produced in various kinds of agricultural farms (Bardi *et al.,* 2013). The issue will get worse unless the farmers change their perception about the adoption of solar energy technologies. Therefore, it is advised for farmers to employ energy-saving tools for producing crops and to maximize land by installing solar energy systems because they are practical, affordable, and useful for a variety of tasks (Duprazet *al.,* 2011). Since many people still solely rely on fossil fuels for energy and lack access to modern energy sources, in light of this, the current study was done to assess the level of awareness and adoption of solar energy in agricultural operations.

Materials and Methods

The present study was conducted in the Hisar district of Haryana state. A total of 8 villages were chosen using a random sampling technique viz., Behbalpur, Bugana, Rajli and Ghirai, Dhani Garan, Jewra, Sarsod, and Bichpari from Barwala block of Hisar district. To collect the data fifteen farmers from each selected village were also selected randomly. As a result, 120 farmers were chosen as responders in the current study. The farmers' socio-personal (age, education, land holding), socio-economic (irrigation facilities available, cropping pattern, farming system, crop rotation followed), and communicational characteristics (mass media exposure, extension contact, social media for getting information), as well as their overall awareness and adoption level for use of solar energy for agricultural operation, were considered. The awareness level of farmers about solar energy for the agricultural operation was measured on a 2-point continuum viz., 'Aware'- '1' and 'Not aware'- '0'. Similarly, the level of adoption ('Adopted'- '1' and 'Not adopted'- '0') and the constraints faced in the adoption of solar energy ('Severe'- '1' and 'Not severe'- '0') were also measured. A purposefully created interview schedule that had been pretested before being given to the sampled respondents was used to gather data from the respondents who had been selected for the sample. Appropriate statistical measures were used i.e., mean, frequency, percentage, and rank order were applied to draw meaningful inferences.

Results and Discussion

Profile of selected respondents:A descriptive analysis of the sampled respondents' socio-personal, socio-economic, and communicational characteristics was conducted.The majority of respondents (56.67%) belonged to the middle age (31 to 50 years) group and had educational qualifications of matriculation (28.33%). According to the present study, about one-third (34.17%) of the respondents have land holding above 5 acres up to 10 acres and above seventy per cent of the respondents (73.33%), have both sources of irrigation (canal + submersible pump/ tube well). A majority (90.00%) were doing livestock with agriculture followed by the integrated farming system (18.33%), practicing sole cropping pattern (65.00%), and having cotton–wheat cropping system (68.33%). In the case of communicational characteristics, the newspaper ranked first with a weighted mean score of 2.31 among the mass media exposure, the progressive farmers with a weighted mean score of 2.65 among the extension contact of the farmers, and the majority got information through WhatsApp (77.50%).

Farmers' awareness about solar energy for agricultural operations: Awareness is a crucial decision-making tool for making farming more profitable and sustainable. Table 1 shows the extent of awareness about solar energy for agricultural operations.

More than half of the respondents were aware about solar energy is an eco-friendly green energy source (68.33%) followed by 'no harmful effect of solar energy on humans and animals (67.50%)', 66.67 per cent respondents were aware about solar energy has the potential of replacing huge usage of diesel used for lighting and irrigation in rural areas, and 65.83 per cent aware about solar energy doesn't need extra land. They are therefore more conscious of environmental deterioration and are as a result taking part in eco-friendly activities and green practices. About 61.67 per cent were aware about solar energy improves the quality of life of both, humans and animals, while 60.00 per cent were aware that solar energy helps to reduce input costs, also about 55.83 per cent were aware that solar pump is the reliable source of energy for irrigation water in a remote area, whereas 55.00 per cent respondents were aware about solar energy gives extra return by supplying to the national power grid and solar energy helps to adopt a customer-centric approach, and 52.50 per cent of respondents aware about solar energy gives extra employment. This shows thatfarmers favour solar energy goods since they have a few extra advantages over traditional ones and they also promote independence. About 51.67 per cent of them were aware about the government's focus on marginal farmers by providing smaller solar pumps through capital subsidy, and solar energy reduces the burden of electricity subsidies on the government. It is clear that they know government support, subsidies, and financial assistance are there for adoption in any agricultural operations. However, 51.67 per cent were aware about solar drying increase the life span of fruits and vegetables for the off-season as well as solar energy helps farmers to fit in all types of cropping system, and 50.83 per cent of total respondents were aware about solar energy helps farmers to increase income from farming by reducing irrigation costs. They know that solar energy will help with respect to production, and postharvest, and increase their income.

On the other hand, only 48.00 per cent were aware about heating & cooling extend the selves life of farm produce, 40.83 per cent were aware about solar energy can be used for pest control also, and only 40.00 per cent of the respondents were not aware about solar energy is an economic, cost-effective, and reliable source and the majority (60.00%) were not aware. It is noted that only 38.33 per cent were aware about it increases agricultural productivity because usage of solar energy when it is needed and about 61.67 per cent were actually not aware of the benefit of its usage. However, only 37.50 per cent were aware that it can increase self-reliance, 36.67 per cent were aware about it can be used as electric fencing of the farm, 35.83 per cent were aware solar energy helps in electricity production for different farm needs/operations, only 35.00 per cent were aware about solar powered pumps restrict greenhouse emission, less pollution as compared to diesel pumps. Whereas only 34.17 per cent were aware and 65.83 per cent of respondents were not aware about it can be suitable for away from the power grid. Further, only 32.50 per cent were aware about no fear of depletion of solar energy as compared to other sources and solar energy helps in the joint ownership model of solar pumps. Whereas only 31.67 per cent of the sampled respondents were aware about solar energy can be used as greenhouse space heating and the majority 68.33 per cent were not aware.

Table 1: Awareness level of farmers about solar energy for agricultural operation (n = 120)

Sr. no.	Statements	Awareness level			
		Aware		Not Aware	
		Frequency	%	Frequency	%
1	Aware about solar energy is eco-friendly green energy source?	82	68.33	28	23.33
2	No harmful effect of solar energy on human and animals	81	67.50	39	32.50
3	Solar energy has a potential of replacing huge usage of diesel used for lighting and irrigation in rural areas	80	66.67	40	33.33
4	Solar energy doesn't need extra land	79	65.83	41	34.17
5	Solar energy improves the quality of life of both, human and animals	74	61.67	46	38.33
6	Solar energy helps to reduce the input costs.	72	60.00	48	40.00
7	Solar energy gives extra return by supplying to nation power grid	66	55.00	54	45.00
8	Solar energy gives extra employment	63	52.50	57	47.50
9	Government focus on marginal farmers by providing smaller solar pumps through capital subsidy	62	51.67	58	48.33

Sr. no.	Statements	Awareness level			
		Aware		Not Aware	
		Frequency	%	Frequency	%
10	Solar energy helps farmers to increase income from farming by reducing irrigation costs	61	50.83	59	49.17
11	Aware that subsidy is provided on solar energy system	57	47.50	63	52.50
12	Solar energy needs extra expertise.	54	45.00	66	55.00
13	Solar energy reduces the burden of electricity subsidies on government.	62	51.67	58	48.33
14	Solar energy improves resilience of farmers against climate change	48	40.00	72	60.00
15	Solar energy helps to adopt a customer-centric approach	66	55.00	54	45.00
16	Solar energy helps in joint ownership model of solar pumps	39	32.50	81	67.50
17	Solar energy helps farmers to fit in all types of cropping system	62	51.67	58	48.33
18	Solar pump is the reliable source of energy for irrigation water in remote area	67	55.83	33	44.17
19	Solar pump also increases, stabilize, and diversified production	53	52.50	37	47.50
20	Solar-powered pumps restrict greenhouse emission, less pollution as compared to diesel pumps	42	35.00	78	65.00
21	Easy to install a solar-powered pump with minimum maintenance cost	47	39.17	73	60.83
22	It can increase the self-reliance	45	37.50	75	62.50
23	Electricity production for different farm needs/operations	43	35.83	77	64.17
24	Most productive method of reducing carbon footprints	61	50.83	59	49.17
25	No fear of depletion as compared to other sources	39	32.50	81	67.50
26	Suitable for away from power grid	41	34.17	79	65.83
27	Solar energy can be used for pest control also	49	40.83	71	59.17
28	Solar drying of vegetables, fruits, and food preservation	60	50.00	60	50.00
29	Solar drying increases the life span of fruits and vegetables for off-season	62	51.67	58	48.33

Sr. no.	Statements	Awareness level			
		Aware		Not Aware	
		Frequency	%	Frequency	%
30	Heating & Cooling extend the selves life of farm produce	58	48.33	62	51.67
31	Solar energy is an economic, cost-effective, and reliable source	48	40.00	72	60.00
32	It increases agricultural productivity because usage of solar energy when it is needed	46	38.33	74	61.67
33	Used as electric fencing for farm	44	36.67	76	63.33
34	Used as lighting of agricultural building/farms	60	50.00	60	50.00
35	Used as greenhouse space heating	38	31.67	82	68.33

As depicted in Table 1, the awareness level is found to be better in the case of solar energy for agricultural operations at various levels, viz., about solar energy is an eco-friendly green energy source (68.33%), no harmful effect of solar energy on human and animals (67.50%), has a potential for lighting and irrigation in rural areas (66.67%), doesn't need extra land (65.83%), solar energy improves the quality of life of both, human and animals (61.67%), and solar energy helps to reduce the input costs (60.00). The overall awareness level is found to be relatively lower in the case of solar energy can be used as greenhouse space heating (31.67%), with no fear of depletion and helps in joint ownership model of solar pumps (32.50%), suitable for away from power grid (34.17%), restrict greenhouse emission (35.00%), and electricity production for different farm needs/operations (35.83%).

Farmers' adoption level about solar energy for agricultural operations: Since adoption is a sign that a technology is practical and widely used, adoption is the primary goal of technology development. The level of adoption of the sampled farmers was measured with the help of an interview schedule. The extent of the adoption of solar energy for agricultural operations is presented in Table 2.

Only 26.67 per cent of the sampled respondents had adopted solar energy because it helps energy saving around the year and the rest of the respondents (73.33%) have not adopted it. Followed by, only 25.00 per cent of respondents had adopted it, as it is a renewable source for agri-production systems and allied enterprises and 75.00 per cent have not adopted the renewable source. Whereas the majority 76.67 per cent of the sampled respondents have not adopted solar energy help in the adoption of suitable crop rotation, only 23.33 per cent had adopted it in their fields. About 77.50 per cent of farmers have not adopted solar energy as the sensible solution to mitigate drought-related problems and solar energy is used for electric fencing of farms to save crops from stray animals. Thus, only 22.50 per cent had adopted it to mitigate drought-based problems and for electric fencing. Only 21.67 per cent of them adopted solar energy as it helps in future cropping patterns depending on energy and adopted solar-powered pumps, which increase the diversified system of production, then about 78.33 per cent have not adopted it for those purposes. Only 20.83 per cent had adopted solar power, as it is an employment creation venture and 79.17 per cent have not adopted it.

Table 2: Adoption level of farmers about use of solar energy for agricultural operation (n = 120)

S. No.	Statements	Adoption level			
		Adopted		Not adopted	
		Frequency	%	Frequency	%
1.	Adoption of solar energy helps energy saving round the year	32	26.67	88	73.33
2.	Solar energy helps in adoption of suitable crop rotation	28	23.33	92	76.67
3.	Solar energy helps in future cropping pattern depends on energy	26	21.67	94	78.33
4.	Solar energy helps to plan the irrigation schedule of crops	20	16.67	100	83.33
5.	Solar energy helps in saving of energy as fuel (diesel and electricity)	22	18.33	98	81.67
6.	Solar energy adopted for irrigation as and when required	16	13.33	104	86.67
7.	Adoption of solar energy helps to reduce all type of pollution	14	11.67	106	88.33
8.	Adoption of solar energy help in timely management of field operation	15	12.50	105	87.50
9.	Adoption of solar energy helps to increase farm income because of availability of energy round the year	15	12.50	105	87.50
10.	Solar energy saves life as there is no harmful effects on human and animals	20	16.67	100	83.33
11.	Solar energy helps in planning of different farm operations	12	10.00	108	90.00
12.	Solar powered pumps also increase the diversified system of production	26	21.67	94	78.33
13.	Solar powered pumps are cost effective reliable, eco-friendly and also offers the self-dependency	23	19.17	97	80.83
14.	Solar powered pumps can be installed in all geographical conditions	22	18.33	98	81.67

S. No.	Statements	Adoption level			
		Adopted		Not adopted	
		Frequency	%	Frequency	%
15.	Solar powered pumps have huge potential to install in remote areas	21	17.50	99	82.50
16.	Solar energy helps agriculture sector around the world to reduce problem of power shortage and fuel costs	22	18.33	98	81.67
17.	Solar energy used as green house space heating	19	15.83	101	84.17
18.	Solar water pump system has an inbuilt inverter which ensure water supply for 24 hours	23	19.17	97	80.83
19.	Solar power is employment creation venture	25	20.83	95	79.17
20.	Solar energy has sensible solution to mitigate drought related problems	27	22.50	93	77.50
21.	Solar energy is a renewable source for agri production system and allied enterprises	30	25.00	90	75.00
22	Solar energy used for electric fencing of farm to save crops from stray animals	27	22.50	93	77.50
Mean adoption score = 4.04					
Overall adoption level = 18.36%					

More than 80 per cent of the sampled respondents have not adopted solar energy, that is only 19.17 per cent had adopted solar water pump system which has an inbuilt inverter that ensures the water supply for 24 hours and solar-powered pumps as it is cost-effective, reliable, eco-friendly and also offers the self-dependency. Whereas only 18.33 per cent of them had adopted solar-powered pumps because they can be installed in all geographical conditions and also help the agriculture sector around the world to reduce the problem of power shortage and fuel costs, therefore it is clear that the majority (81.67%) have not adopted it in the fields. However, solar-powered pumps have huge potential to install in remote areas, only 17.50 per cent of the respondents had adopted it. Furthermore, solar energy helps to plan the irrigation schedule of crops and solar energy saves life as there are no harmful effects on humans and animals, only 16.67 per cent of them had adopted and 83.33 per cent have not adopted it. Only 15.83 per cent of respondents had adopted solar energy as a greenhouse space heating. Similarly, only 13.33, 12.50, 11.67 per cent, adopted solar energy for irrigation as and when required, helps to increase farm income because of the availability of energy around the year and help in timely management of field operation, helps to reduce all type of pollution, respectively at their farm fields. It is noted that about 90.00 per cent of the sampled respondents have not adopted solar energy help in the planning of different farm operations and only 10.00 per cent had adopted it. The overall adoption level of farmers found as 18.36 per cent, with a mean adoption score of 4.04.

As depicted in Table 2, the adoption level is found to be relatively lower in almost all cases i.e., the level of adoption of solar energy was only to the extent of 26.67, 25.00, 23.33, 22.50, 21.67, 20.83, 19.17, 18.33, 16.67, 13.33, 12.50, 11.67 and 10.00 per cent, respectively at farmers' fields. As for as, the non-adopted category of respondents about solar energy used in agricultural operations in the sequence as stated above i.e., 73.33, 75.00, 76.67, 78.33, 80.83, 81.67, 82.50, 83.33, 84.17, 86.67, 87.50, 88.33, and 90.00 per cent, respectively the respondents not adopted any solar energy for agricultural operations at their fields.

Therefore, the present study revealed overall adoption of solar energy for agricultural operations is at a low level (18.36%). According to Saini and Devi's study from 2019, the majority of people (58%) are aware of the fundamentals of solar energy, and 50% of respondents are knowledgeable about the technical aspects of solar energy systems, such as cost and capacity. Additionally, 56.5% of respondents are aware of government initiatives aimed at promoting solar energy, while 40.3% of respondents are unaware of it.According to Ibrahim and Kumari (2020), utilizing solar technology in farming has many benefits, including lowering the cost of agricultural production, reducing climate change, and creating rural jobs. A study by Kumar *et al.,* (2020) revealed that in the non-adoption of solar-powered pumps, the majority of respondents agreed that investment cost and maintenance cost are the two factors that have the biggest impact. In their research on Bangladesh, Muhammad, Mosleh, and Hague Khan (2014) stated that the utilization of solar irrigation systems had a significant economic advantage over conventional grid-electricity-driven pump irrigation systems or fuel-based irrigation, and also increased farmer yields, lower the cost of agricultural production. The findings of Upadhyay and Chowdhury (2014) stated that the adoption process is largely impacted by the price of solar energy-based products. Bandara and Amarasena (2020) study findings indicate that perceived cost has a statistically significant detrimental impact on solar energy uptake. In many ways, these findings are comparable to those of the current study.

Farmers' constraints faced in the adoption of solar energy: There are various problems/constraints faced by the farmers in the adoption of solar energy and measured with the sampled respondents using the developed interview schedule. Table 3 shows the extent of constraints faced in the adoption of solar energy for agricultural operations.

Table 3: Constraints faced by farmers in the adoption of solar energy (n = 120)

S. No.	Constraints	Severe (1)	%	Not Severe (0)	%
1	It is not an absolutely reliable source of energy on cloudy days	104	86.67	06	13.33
2	Very high initial installation costs	102	85.00	08	15.00
3	Solar energy storage is very expensive	101	84.17	19	15.83
4	Micro cracks are a common issue with solar panel	98	81.67	22	18.33
5	Solar energy is weather dependent	97	80.83	23	19.17
6	Not efficient where groundwater is very deep	96	80.00	24	20.00
7	Low energy conservation rates	94	78.33	26	21.67
8	Not suitable for intensive cropping pattern	93	77.50	27	22.50
9	Small and marginal farmers are not familiar with this system	92	76.67	18	23.33
10	Complications when moving the system	90	75.00	30	25.00
11	Solar panels and fencing are costly	88	73.33	32	26.67
12	Solar water pump installation is a herculean task	85	70.83	35	29.17
13	Less efficient in sprinkler system	74	61.67	46	38.33
14	Fear of theft of solar plates in remote areas	72	60.00	48	40.00
15	Problems of efficiency of the solar system	71	59.17	09	49.17

The majority 86.67 per cent of respondents felt that solar energy is not an absolutely reliable source of energy on cloudy days, and about 85.00 per cent felt that it involves very high initial installation costs. Similarly, 84.17 per cent of the respondents felt that solar energy storage is very expensive, which might be the major reason for the less adoption. Constraint like micro-cracks on a solar panel was felt by 81.67 per cent of the farmers, and 80.83 per cent of them felt solar energy is weather dependent as the constraint they faced in the adoption of it. Whereas 80.00 per cent felt that it is not efficient where groundwater is very deep, 78.33 per cent felt that solar energy has low energy conservation rates, and 77.50 per cent of them felt that it is not suitable for intensive cropping patterns. However, 76.67 per cent faced problems because small and marginal farmers are not familiar with this system to be utilized in their fields. Whereas 75.00 per cent felt complications when moving the system, so it declines its adoption. Moreover, 73.33 per cent felt solar panels and fencing are costly, and 70.83 per cent of the sampled respondents felt solar water pump installation is a herculean task. More than half of the respondents (61.67%) felt solar energy is less efficient in sprinkler systems, 60.00 per cent of them felt fear of theft of solar plates in remote areas, and 59.17 per cent felt problems with the efficiency of the solar system, respectively, as mentioned by the respondents.

As depicted in Table 3, the major constraints found in the case of solar energy adoption for agricultural operations at various levels, viz., solar energy is not an absolutely reliable source of energy on cloudy days (86.67%), very high initial installation costs (85.00%), storage is very expensive (84.17%), micro-cracks on a solar panel (81.67%), is weather dependent (80.83%), not efficient where groundwater is very deep and low energy conservation rates by 80.00, and 78.33 per cent farmers, respectively.

Conclusion

Globally, governments, businesses, and other public and private organizations are increasingly seeking solutions for lowering the emissions of greenhouse gases from routine industrial and mechanical activities by shifting to green energy or transitioning to long-term renewable energy sources like solar energy. The present study has revealed an overall gap of about 18 per cent between farmers' awareness and adoption level of solar energy for agricultural operations which needs to be bridged. The awareness of the farmers varied across solar energy adoption for agricultural operations. Awareness level was favourable with more than 50 per cent level of awareness about solar energy as an eco-friendly green energy source, no harmful effect of solar energy on humans and animals, has the potential for lighting and irrigation in rural areas, and doesn't need extra land, solar energy improves the quality of life of both, human and animals, and solar energy helps to reduce the input costs. However, more effort is needed to increase the adoption level of farmers about the use of solar energy in agricultural operations, which is very low. Farmers require sufficient technological direction like training in the operation and maintenance of solar energy technologies. There must be a solid support service from extension personnel to increase the use of solar energy in agricultural operations. Government policymakers must take efficiency and cost into account while developing solar energy technologies in order to encourage farmers to adopt them, which leads to a path toward green energy use.

References

Aroonsrimorakot, S., Laiphrakpam, M. and Paisantanakij, W. (2020). Application of solar panel energy technology for sustainable agriculture farming. International Journal of Agricultural Technology 16(3): 553-562.

Aroonsrimorakot, S. and Laiphrakpam, M. (2019). Application of solar energy technology in agricultural farming for sustainable development: A review article. International Journal of Agricultural Technology, 15: 685-692.

Bardi, U., El Asmar, T. and Lavacchi, A. (2013). Turning electricity into food: the role of renewable energy in the future of agriculture. Journal of Cleaner Production, 53: 224-231.

Dupraz, C., Marrou, H., Talbot, G., Dufour, L., Nogier, A., & Ferard, Y. (2011). Combining solar photovoltaic panels and food crops for optimising land use: Towards new agrivoltaic schemes. Renewable energy, 36(10): 2725-2732.

Mekhilef, S., Faramarzi, S. Z., Saidur, R. and Salam, Z. (2013). The application of solar technologies for sustainable development of agricultural sector. Renewable and sustainable energy reviews, 18: 583-594.

Muhammad, D., Mosleh, M., and Khan, S. U. H. (2014). Assessment and evaluation of solar irrigation system in Bangladesh. International Conference on the Developments in Renewable Energy Technology (ICDRET), 1-6.

Rimpi Saini and Rajni Devi. (2019). Awareness and Attitude towards Solar energy in Kurukshetra Haryana. International Journal of Scientific Development and Research (IJSDR), 4(4): 176-181.

Torshizi, M. V. and Mighani, A. H. (2017). The application of solar energy in agricultural systems. Journal of Renewable Energy and Sustainable Development, 3(2): 234-240.

Bandara, U. C., and Amarasena, T.S.M. (2020). Impact of Perceived Ease of Use, Awareness and Perceived Cost on Intention to Use Solar Energy Technology in Sri Lanka. Journal of International Business and Management 3(4): 01-13.

Ibrahim, M. S., and Oum Kumari, R. (2020). Emerging Solar Energy Technologies for Sustainable Farming: A Review. Journal of Xi'an University of Architecture and Technology, 12: 5328.

Kumar, V., Syan, A. S., Kaur, A. and Hundal, B. S. (2020). Determinants of farmers' decision to adopt solar powered pumps. International Journal of Energy Sector Management, 14(4): 707-727.

Upadhyay, A. and Chowdhury, A. (2014). Solar Energy Fundamentals and Challenges in Indian restructured power sector. International Journal of Scientific and Research Publications, 4(10): 1-13.

18

Marketing Behaviour of Vegetable Growers in East Singhbhum District of Jharkhand

Abhijeet Satpathy[1*], R.S. Panigrahi[2], Sarbani Das[3] and Harihara Jena[4]

[1]Birsa Agricultural University, Ranchi, Jharkhand
[2,3]Odisha University of Agriculture & Technology, Bhubaneswar
[4]SRF, Biotech-KISAN Project, Bhawanipatna, Odisha

Abstract

The research study was conducted purposively in the East Singhbhum district of Jharkhand with the objectivesto determine the marketing behaviour of vegetable growers and to find out the relationship between the socio-economic attributes with the marketing behaviour of the vegetable growers. The sample size covered 120 vegetable growers from 10 villages & response was obtained from each respondent with the help of a structured interview schedule pretested with 10 percent samples other than the respondents of the study. The research design employed in this study was ex-post-facto. The findings of the study revealed that the majority of the respondents (97.5%) were growing vegetables because of regular income (daily), majority of the respondents (92.5) expressed that no cold storage facilities available in the area were the major reason for selling vegetable at a particular period, majority of the respondents (77.5%) expressed that they sold directly to the consumers in the village market. It was also revealed that majority of the respondents (87.5%) expressed that they sold their produce to a particular agency near the village because they got immediate cash payments through it, most of the respondents (75.8%) sold their produce in the nearby weekly markets, the reason might be better transportingfacilities available at the market and better prices in the market. Majority 43.34 percent of the respondents had a medium level of marketing behaviour followed by 33.33 percent respondents had a low level of marketing behaviour. From the study of correlation between socio-economic attributes and overall marketing behaviour of vegetable growers it was revealed that social participation, extension contact and risk orientationwere positively and significantly related to marketing behaviour of vegetable growers.

Keywords: *Marketing behavior, Vegetable growers, Correlation, Risk orientation, Extension contact*

Introduction

India is the second-largest producer of vegetables in the world after China. India accounts for 16 percent of the world's production of vegetables. India home to 17.74% of the world's population has only accounted for about 2.4% of the world's geographical area and 4% of its water resources. Vegetables play an imperative role in nutritional security, economic viability and fit well into the predominant intensive cropping systems prevailing in different parts of our country. More than 60 kinds of vegetables are grown in India in tropical, subtropical and temperate agro-climates. India is the largest producer of okra and ranks 2nd in the production of potato, onion, cauliflower, and cabbage. During 2013-2014, India produced 162.19 million tonnes of vegetables and vegetable export from our country was worth Rs. 5462.93 crores. For a balanced diet, an adult who works moderately needs about 300 grams of vegetables per day, of which about 50 grams should comprise root vegetables, 50 grams of leafy vegetables and 200 grams of other kinds of vegetables as per Dietary Guidelines for Indians, NIN, 2011.

The vegetable is one of the important crops of Jharkhand, which is being cultivated on a commercial scale. About 3.2% of the gross cropped area of the State is devoted to vegetable crops. Vegetable development depends not only on production but also on the marketing system. Vegetable cultivation being labour intensive can substantially increase employment and livelihood avenues too. Vegetable growers face many challenges in the course of marketing their products and their marketing behaviour varies under different circumstances. Keeping this in view, the present investigation entitled "Marketing Behaviour of vegetable growers in East Singhbhum district of Jharkhand" was undertaken.

Materials & Methods

In this study, an Ex-post-facto research design was used. This design is appropriate because the phenomenon has already occurred. The state of Jharkhand was selected purposively because it has immense potential for the marketing of agricultural produce and establishment of the local market, Hatt / weekly market, etc. East Singhbhum district was selected purposively for the study. This district contributes a major share to the vegetable production of Jharkhand. The study was conducted in Mosaboni, Ghatshila, Baharagora, Dumaria, Patamda Blocks of East Singhbhum district of Jharkhand. Both purposive and random sampling techniques were followed in the study. The districts and blocks were selected purposively, whereas the villages were selected randomly and from ten villages, 120 respondents were selected through a disproportionate random sampling method.The primary data were collected through a personal interview method with the help of a pre-tested interview schedule, which was prepared based on objectives of investigation and variables. The interview schedule was thoroughly discussed with the member of the advisory committee and their suggestions were incorporated. The statistical tests and procedures were used for analyzing the data with the help of statistical tools like- mean, S.D., the percentage was used for the analysis of data.

Results & Discussions

Marketing Behaviour of Vegetable Growers

Marketing behaviour refers to the behaviour of the farmers with respect to marketing aspects of vegetable including reasons for growing vegetables, time of sale, place of sale, and marketing channels used, reasons for selling, etc. The Information regarding reasons for growing vegetables, reasons for selling at a particular period/time, whom the produce is sold, reasons to sell to a particular agency, where the produce is sold and reasons for selling at a particular place were collected and analyzed to get an idea of the marketing behaviour of vegetable growers.

Table 1: Distribution of respondents according to their marketing behaviour

Sl.No.	Category	Respondents	
1	**Various reasons for growing vegetables**	**F**	**%**
a	Highly remunerative than other crops	108	90.0
b	Availability of resources	79	65.8
c	Good market demand for vegetables	110	91.7
d	Govt. encouragement and assistance	40	33.3
e	Marketing facilities	52	43.3
f	Regular income(daily)	117	97.5
2.	**Reasons for selling at a particular period/ time?**		
a	Highly perishable	107	89.2
b	Quality was not good	12	10.0
c	No cold storage facilities available	111	92.5
d	Financial urgency	40	33.3
e	Indebtedness to trader	4	3.3

3.	**Whom do you sell the produce?**		
a	Directly to the consumer	93	77.5
b	To village level Trader/ Baniya	24	20.0
c	To the wholesaler through commission agents	84	70.0
d	To the trader through co-operative society	0	0
e	Retailers	2	1.7
4.	**Reasons to sell to a particular agency**		
a	The agency is very nearer one	25	20.8
b	The agency is worthy credit	14	11.7
c	I have no time to engage myself in selling directly to the consumer	88	73.3
d	Immediate cash payment	105	87.5
e	Better price	98	81.7
5.	**Where do you sell the produce**		
a	In the village	84	70.0
b	In the nearby weekly market	91	75.8
c	In the distant market	27	22.5
6.	**Reasons for selling at a particular place**		
a	The market is very near to the place	80	66.7
b	better transport facilities available are for that place	59	49.2
c	better market facilities available in that market	14	11.7
d	The better prices are available in that market	60	50.0

From the above table it is identified thatthe majority (97.5%) were growing vegetables because of regular income(daily), followed by good market demand for vegetables (91.7%), Highly remunerative than other crops (90%), Availability of resources (65.8%), Marketing facilities (43.3%) and Govt. encouragement and assistance (13.33%) respectively.

Majority of the respondents (92.5) expressed that no cold storage facilities available in the area were the major reason for selling vegetable at a particular period, whereas (89.2) of them sold their produce because of perishable nature of the vegetables followed by financial urgency (33.3), quality was not good (10), and a very few (3.3) due indebtedness to the trader respectively.

Majority of the respondents (77.5%) expressed that they sold directly to the consumers in the village market. Further (70%) sold their produce to the wholesaler through commission agents, the reason might be the immediate cash payment to the vegetable growers followed by (20%) to the village level trader/ baniya and retailers (1.7%). None of them sold their produce to the traders through commission agents.

Majority of the respondents (87.5%) expressed that they sold their produce to a particular agency near the village because they got immediate cash payments through it whereas (81.7%) of the respondents expressed that they sold their produce to a particular agency because they availed better price of the product from the particular agency, (73.3%) of them sold their products because they have no time to engage themselves in selling directly to the consumers, followed by (20.8%) the agency is nearer one and (11.7%) the agency is worthy credit.

Most of the respondents (75.8%) sold their produce in the nearby weekly markets, the reason might be better transporting facilities available at the market and better prices in the market followed by (70%) in the village market and (22.5%) in the distant markets.

Around (66.7%) of the respondents expressed that they sold their produce because of nearness of market followed by (50.0%) due to better prices are available in that market, (49.2%) due to better transport facilities available at the market and very few (11.7%) due to better facilities available at the market.

Overall Marketing Behaviour of the Vegetable Growers

Some of the variables like reasons for growing vegetables, reasons for selling at a particular time, the customer of the product, place of sale, reasons to sell to a particular agency and reasons for selling at a particular place were collected and analyzed.

Table 2: Overall marketing Behaviour of vegetable growers

Sl.No	Marketing Behaviour	Frequency	Percentage
1	Low (Mean- 1/2 S.D)	40	33.33
2	Medium (Mean ± 1/2 S.D)	52	43.34
3	High (Mean + 1/2 S.D)	28	23.33
	Total	120	100

Mean = 43.38, S.D.= 1.97

It is clear from the table that the majority 43.34 percent of the respondents had a medium level of marketing behaviour followed by 33.33 percent respondents who had a low level of marketing behaviour and only 23.33 percent of the respondent had a high level of marketing behavior.

Correlation Between Socio-economic Attributes and Marketing Behaviour of Vegetable Growers.

The relationship between the socio-economic attributes with the marketing behaviour of the vegetable growers. In order to study the nature of the relationship between the selected independent variables and the marketing behaviour, correlation coefficient (r) was computed and the findings are described below in the table below

Table 3: Correlation between socio-economic attributes and marketing behaviour of vegetable growers (N=120)

Variables	Reasons for growing vegetables	Reasons for selling at a particular time	Customer of the produce	Place of sale	Reasons to sell to agency	Reasons for selling at a particular place	Overall marketing behaviour
Age	0.205*	0.133*N*	-0.174*N*	-0.137*N*	0.073*N*	-0.165*N*	0.46*N*
Education	0.076*N*	-0.169*N*	-0.027*N*	-0.043*N*	-0.346**	-0.145*N*	-0.239**
Housing pattern	0.312**	-0.153*N*	-0.261**	-0.150*N*	-0.099*N*	0.118*N*	0.008*N*
Landholding	0.154*N*	-0.080*N*	-0.249**	-0.007*N*	0.101*N*	0.078*N*	0.052*N*
Occupation	-0.216*	0.184*	0.024*N*	0.034*N*	-0.066*N*	0.106*N*	-0.041*N*
Income	0.328**	-0.012*N*	-0.087*N*	-0.103*N*	0.102*N*	-0.019*N*	0.178*N*
Social participation	0.437**	-0.068*N*	-0.185*	-0.311**	0.127*N*	0.119*N*	0.194*
Extension contact	0.488**	-0.071*N*	-0.137*N*	-0.233**	0.098*N*	0.125*N*	0.254**
Source of information	0.541**	-0.079*N*	-0.355**	-0.334**	-0.001*N*	0.086*N*	0.115*N*
Risk orientation	0.510**	-0.109*N*	-0.137*N*	-0.362**	0.272**	0.042*N*	0.261**
innovativeness	0.274**	-0.094*N*	-0.140*N*	-0.337**	0.000	-0.234*	-0.097*N*

*Significant at 5% level of significance

**Significant at 1% level of significance

N – Not significant

A correlation between social attributes and overall marketing behaviour of vegetable growers in table above revealed that:

The “r” value of Education and the overall marketing behaviour was -0.239**, being significant at 0.01 level of significance. Thus, it is concluded that education has shown negative significant relationship with reasons. Hence null hypothesis was rejected.

The “r” value of social participation and the overall marketing behaviour was 0.194*, being significant at 0.05 level of significance. Thus, is concluded that social participation has shown a significant relationship with overall marketing behaviour. Hence null hypothesis was rejected.

The “r” value of Extension contact and the overall marketing behaviour was 0.254**, being significant at 0.01 level of significance. Thus, it is concluded that extension contact has shown a significant relationship with overall marketing behaviour. Hence null hypothesis was rejected

The “r” value of Risk orientation and the overall marketing behaviour was 0.261**, being significant at 0.01 level of probability. Thus, it is concluded that risk orientation has shown a significant relationship with overall marketing behaviour. Hence null hypothesis was rejected.

Conclusion

The study revealed that the vegetable growers cultivate various types of vegetables based on their feasibility.During the cultivation of vegetables, they faced many challenges ranging from the distant location, distress sale and high commission of middlemen. The study indicated that the majority of the respondents were selling their produce daily, the majority of the respondents expressed that no cold storage facility was available in the area, majority of the respondents expressed that they sold their produce directly to the consumer, most of them expressed that they sold their produce to a particular agency because they get immediate cash payment for their produce, majority of the respondents sold their produce in the nearby weekly market because of the nearness of the market.

References

Balusu, K. B. 2014. Marketing behavior of vegetable growers in Ranga Reddy district of Andhra Pradesh, Thesis, Acharya N.G. Ranga Agricultural University, Hyderabad, Hyderabad.

Barik, J. 2013. A study on the entrepreneurial behavior of vegetable growers of Cuttack district, M.Sc. (Ag.)Thesis, Orissa University of Agriculture and Technology.

Johnson, B. and Manoharan, B. (2009). Marketing Behaviour of Cashew Farmers. Indian Res. J. Ext. Edu., 9 (1): 6-10

Madhu Shekar, B. R. 2009, Study on Marketing Behaviour of Chilli Growers In Guntur District Of Andhra Pradesh, M.Sc. (Ag.) Thesis, Acharya N.G. Ranga Agricultural University, Hyderabad

Mohapatra, S.C., (1999). Production and marketing of onion in Bolangir district of Orissa. *Indian J. Agric. Markt.*,**13**(1): 40-43.

Seetharaman, R., Netaji, Muthaiah. S., Velusamy, R. 2001. Marketing behavior of flower cultivators, Agricultural Marketing Interventions and Innovations, National Institute of Agricultural Extension Management, Hyderabad,: 126- 141.

19

Entrepreneurial Behaviour of Turmeric Growers

Dhande S.P.[1], Manisha L.D.[2*], Mahajan S.A.[3] and Mandve R.P.[4]

[1,3,4]Section of Extension Education, College of Agriculture Nagpur, Maharashtra
[2]Department of Extension Education, Post Graduate Institute Dr. PDKV, Akola, Maharashtra

Abstract

The present study on "Entrepreneurial behaviour of turmeric growers" was conducted in Vidarbha region of Maharashtra with sample size of 120 farmers from 15 villages. An exploratory research design was used for the investigation. Data was collected by personally interview schedule. The data collected was carefully examined, classified, quantified and tabulated. Frequencies, percentage, mean, standard deviation and correlation were employed for interpreting the results. In case of the entrepreneurship behaviour of turmeric growers by considering different parameters like decision making ability, Risk bearing ability, achievement motivation and knowledge of the enterprise, majority of respondents (73.33%) of the turmeric growers had medium level of entrepreneurial behaviour followed by high (14.17%) and low (12.50%), respectively. The findings of the research revealed that majority of the farmers had medium entrepreneurial behaviour, which indicates that the turmeric growers are moving towards progress and for transformation of the responded from the medium to high entrepreneurial behaviour, it need support from the government agencies to develop strategies for the entrepreneurs to maximize related income generation of the turmeric growers through turmeric enterprise. There is need to provide advanced agriculture technologies to the turmeric growers and motivate them to adopt this technologies.

Keywords: Turmeric cultivation, entrepreneurial behaviour, respondents, Vidarbha, Implication

Introduction

India is the largest producer, consumer and exporter of turmeric in the world. Indian turmeric is considered to be the best in the world market because of its high curcumin content. India accounts for about 80 per cent of world turmeric production and 60 per cent of world exports. Major turmeric exporting countries are India, Thailand, Taiwan, and several other Southeast Asian, Central and Latin American countries. The major turmeric importing countries include Japan, Sri Lanka, Iran, UAE, US, UK and Ethiopia.

Turmeric is native of India and china. The word turmeric is derived from the French word 'Terre-merite' meaning merit of the earth. Its botanical name is Curcuma longa L. and belongs to family Zingeberaceae, it is also known as 'Indian Saffron' is an herbaceous perennial plant growing up to the height of 60 – 90 cm with short stem. The plant is propagated from rhizomes. The leaves are long, broad lanceolate and bright green. The flowers are pale yellow and born on dense spikes. The pseudo stem are shorter than leaves. The rhizomes were ready for harvesting about 7 to 9 months after planting.

India is popularly known as the "Spiced Bowl of the World" as a wide variety of spices with premium quality grown in a country since ancient times. Indian turmeric is considered to be the best in the world due to presence of high curcumine content. Turmeric contains up to 5.00 per cent essential oil and up to 5.00 per cent curcumine, a polyphenol.

As a medicine turmeric has been used in Ayurvedic system of medicine in India from times immemorial. It is claimed to be a stomaching tonic, blood purified, antiseptic, antacid, antiperiodic and carminative. Turmeric is very important spice in India. It is also known as Indian saffron. In India companies like Godrej, vicco are engaged in manufacturing different new products from turmeric like soaps, antiseptics, cosmetics, etc. Turmeric or Haladi powder boiled in milk along with small quantity of black paper may be taken two-three times a day for sore throat, cough, cold and other acute respiratory infection.

Turmeric occupied 6% of the total area under spices and condiments in India. Erode, a city in the South Indian state of Tamil Nadu, is the world's largest producer of and the most important trading centre for turmeric. It is also known as "Yellow City," or "Turmeric City," Sangli, a city of Maharashtra, is second only to Erode in size and importance as a production and trading site for turmeric.

Maharashtra state in India ranks sixth in area under turmeric cultivation. The area under turmeric crop was 11000 hectares with a production of 45000 tonnes and productivity of 4.09 tonnes/hectare during 2015-16 (A. J. Kadte et al). In Maharashtra Sangali, Satara, Hingoli, Nanded, Parbhani are the major turmeric growing districts. In selected Wardha district for research, area under cultivation in 2018-2019 is 590.63 ha (Anonymous 2019). The Waigaon area comes under Wardha their variety knows as Waigaon turmeric recently got a geographical indication (GI) tag. Turmeric has a great importance as spice and condiments. It is useful for industrial purpose.

Entrepreneurial behaviour of farmers is operationally defined as cumulative outcome of several components namely decision making ability, risk bearing ability, achievement motivation, knowledge of the enterprise and managerial assistance. A farmer with good entrepreneurial qualities will make use of his potential in taking up of different activities relating to production and marketing aspects of Turmeric cultivation, ultimately reflects his better farming.

The entrepreneurship is the vital element of any country for promoting economic growth and technological change. The development of entrepreneurship is directly related to the socioeconomic development of the society. Eighty per cent of the population resides in the rural areas and seventy per cent of the work force depends on agriculture for their livelihood. Therefore, role of farmers is very important in agricultural and socio-economic development of the nation.

This study is undertaken to find impact of entrepreneurial behaviour on farming of turmeric growers in Wardha district of Maharashtra. Besides this, the findings of the study would reveal the profile characteristics of the respondents and their entrepreneurial behaviour in turmeric cultivation which may be of great help to the extension workers in formulating different strategies suited to different clientele in turmeric farming.

A probe into the entrepreneurial behaviour in turmeric farming, would serve as indicators for the state department officials, policy makers and extension administrators to plan their future programmes efficiently and make suitable changes in the programmes so as to make the turmeric farming more sustainable and profitable.

Methodology

The present study was carrying out in Wardha district of Maharashtra State. In Wardha district there are 8 tahsils, out of which five tahsil namely, Samudrapur, Hinganghat, Wardha, Deoli and Arvi were selected on the basis of number of respondents cover. From each of selected tahsils 3 villages were randomly selected, thus 15 villages were selected for the study. From all the selected villages 8 farmer from each village were selected randomly and was treated them as respondents under present study, Hence total 120 respondents was selected for the study. Exploratory design of social research was used. The data were collected in face-to-face situation by the personal interview method with the help of structured interview schedule containing the questions onentrepreneurial behaviour of turmeric growers with considering decision making ability, risk bearing ability, achievement motivation, knowledge of the enterprise and managerial assistance.

Result

The findings of the study has presented under the following headings,

1. Components of Entrepreneurial Behaviour

1.1 Decision Making Ability

Table 1: Distribution of respondents according to their decision making ability

Sl.No	Decision criteria	Not considered F (%)	Consider after consulting others F (%)	Considered independently F (%)
1	Increase or decrease profits of the turmeric enterprise	0 (0.00)	5 (4.17)	115 (95.83)
2	Borrow money for the turmeric enterprise	105 (87.50)	7 (5.83)	8 (6.67)
3	Try new income generating activities	9 (7.50)	66 (55.00)	45 (37.50)
4	Switch to new ideas or economic activities	6 (5.00)	65 (54.17)	49 (40.83)
5	Hire workers in the turmeric enterprise	3 (2.50)	5 (4.17)	112 (93.33)
6	Attend training, demonstrations with respect of turmeric etc.	3 (2.50)	69 (57.50)	48 (40.00)
7	To buy new equipment needed for turmeric enterprise	97 (80.83)	8 (6.67)	15 (12.50)

Table 1 indicated that, majority 95.83 per cent of the respondents had considered independently about Increase or decrease profits of the turmeric enterprise followed by 4.17 per cent respondents consider after consulting others. Majority 87.50 per cent of respondents not considered about borrow money for the turmeric enterprise, followed by 6.67 per cent and 5.83 per cent of the respondents considered independently and consider after consulting others, respectively. Majority 55.00 per cent of the respondents consider after consulting others try new income generating activities, followed by 37.50 per cent and 7.50 per cent of respondents considered independently and not considered, respectively. Majority 54.17 per cent of the respondents considered after consulting others switch to new ideas or economic activities, followed by 40.83 per cent respondents considered independently and 5.00 per cent of respondents not considered. Majority 93.33 per cent of the respondents considered independently hire workers in the turmeric enterprise, followed by 4.17 per cent and 2.50 per cent of respondents consider after consulting others and not considered respectively. Majority 57.50 per cent of the respondents considered after consulting others attend training, demonstrations with respect of turmeric etc., followed by 40.00 per cent and 2.50 per cent of respondents considered independently and not considered, respectively. Majority 80.83 per cent of the respondents not considered to buy new equipment needed for turmeric enterprise, followed by 12.50 per cent and 6.67 per cent of respondents considered independently and considered after consulting others, respectively.

Table 2: Distribution of respondents according to their decision making ability index

Sl.No.	Categories	Frequency	Percentage
1	Low (up to 43.53)	14	11.67
2	Medium (43.54 to 78.61)	92	76.66
3	High (above 78.62)	14	11.67
	Total	120	100
	Mean=61.07		

Table 2 indicated that, majority 76.66 per cent of respondents had medium decision making ability, whereas, low 11.67 per cent and high 11.67 per cent of respondents had decision making ability of turmeric growers.These results are in conformity with Bindu (2019).

Risk Bearing Ability

Table 3: Distribution of respondents according to their risk bearing ability

Sl.No.	Statement	Yes F (%)	No F (%)
1	Adoption of any new enterprise for turmeric the first time in their village	41 (34.17)	79 (65.83)
2	Stop the growing turmeric as an enterprise due to the losses.	22 (18.33)	98 (81.67)
3	Use of any modern equipment in your turmeric enterprise	110 (91.67)	10 (8.33)
4	Taken a loan for purchasing equipment needed for your turmeric enterprise	10 (8.33)	110 (91.67)
5	Try to any new turmeric growing practices in your enterprise by hearing to radio/ watching TV / from relatives	61 (50.83)	59 (49.17)

Table 3 indicated that, majority 65.83 per cent of the respondents had never adopted of any new enterprise for turmeric the first time in their village and 34.17 per cent of respondents had adopted of any new enterprise for the first time in their village. The majority 81.67 per cent of the respondents had never stop the growing turmeric as an enterprise due to the losses and 18.33 per cent of the respondents had stop the growing turmeric as an enterprise due to the losses. Majority 91.67 per cent of the respondents used of any modern equipment in their turmeric enterprise and only 8.33 per cent of respondents had not used the any modern equipment in their turmeric enterprise. Majority 91.67 per cent of the respondents had never taken a loan for purchasing equipment needed for their turmeric enterprise and 8.33 per cent of the respondents had taken a loan for purchasing equipment needed for their turmeric enterprise. Majority 50.83 per cent of the respondents had try to any new turmeric growing practices in their enterprise by hearing to radio/ watching TV / from relatives and 49.17 per cent of respondents had not try to any new turmeric growing practices in their enterprise by hearing to radio/ watching TV / from relatives.

Table 4: Distribution of respondents according to risk bearing ability index

Sl.No.	Category	Frequency	Percentage
1	Low (up to 22.24)	30	25.00
2	Medium (22.25 to 60.76)	77	64.17
3	High (above 60.77)	13	10.83
	Total	120	100
	Mean=41.50		

Table 4 indicated that majority of 64.17 per cent of respondents had medium risk bearing ability, whereas, low 25.00 per cent and high 10.00 per cent of respondents had risk bearing ability of turmeric growers.These results are in conformity with jenila (2018) and kolghane (2018).

Achievement Motivation

Table 5: Distribution of respondents according to their achievement motivation.

Sl.No.	Statement	SA F (%)	A F (%)	UD F (%)	DA F (%)	SDA F (%)
1	Work should come first even if one cannot get proper rest in order to achieve ones goals.	8 (6.67)	110 (91.66)	2 (1.67)	0 (0.00)	0 (0.00)
2	It is better to be content with whatever little one has, than to be always struggling for more.	4 (3.33)	38 (31.67))	28 (23.33)	49 (40.84)	1 (0.83)
3	No matter what I have done I always want to do more.	. 9 (7.50)	94 (78.33)	17 (14.17)	0 (0.00)	0 (0.00)
4	I would like to give up at something which is difficult to achieve.	3 (2.50)	92 (76.67)	21 (17.50)	4 (3.33)	0 (0.00)

Sl.No.	Statement	SA F (%)	A F (%)	UD F (%)	DA F (%)	SDA F (%)
5	The way things are happening now a days discourage one to work hard.	3 (2.50)	68 (56.67)	37 (30.83)	12 (10.00)	0 (0.00)
6	One should have determination and driving ambition, even if these qualities make him unpopular.	7 (5.83)	109 (90.84)	3 (2.50)	1 (0.83)	0 (0.00)

Table 5 indicated that, majority 91.66 per cent of the respondents had agree with work should come first even if one cannot get proper rest in order to achieve ones goals followed by 6.67 per cent and 1.67 per cent of the respondents had strongly agree and undecided respectively. More than 40.84 per cent of the respondents had disagree with it is better to be content with whatever little one has, than to be always struggling for more followed by agree 31.67 per cent, undecided 23.33 per cent, strongly agree 3.33 per cent and strongly disagree 0.83 per cent. Majority 78.33 per cent of the respondents had agree with no matter what I have done I always want to do more followed by 14.17 per cent and 7.50 per cent of the respondents had undecided and strongly agree respectively. Majority 76.67 per cent of the respondents had agree with I would like to give up at something which is difficult to achieve followed by undecided 17.50 per cent, disagree 3.33 per cent and strongly agree 2.50 per cent. Majority 56.67 per cent of the respondents had agree with the way things are happening now a day's discourage one to work hard followed by undecided 30.83 per cent, disagree 10.00per cent and strongly agree 2.50 per cent. Majority 90.84 per cent of the respondents agree with one should have determination and driving ambition, even if these qualities make him unpopular followed by strongly agree 5.83 per cent, undecided 2.50 per cent and disagree 0.83 per cent.

Table 6: Distribution of respondents according to their achievement motivation index

Sl.No.	Category	Frequency	Percentage
1	Low (up to 53.25)	13	10.83
2	Medium (53.26 to 79.70)	92	76.67
3	High (above 79.71)	15	12.50
	Total	120	100
	Mean=66.47		

Table 6 indicated that, majority 76.67 per cent of respondents had medium achievement motivation, whereas, high 12.50 per cent and low10.83 per cent of respondents had achievement motivation of turmeric growers.These results are in conformity with Wankhade *et al.* (2013) and Kolgane *et al.* (2018).

Knowledge of the Enterprise

Table 7: Distribution of respondents according to their knowledge of the enterprise

Sl.No.	Statement	Yes F (%)	No F (%)
1	Modern technology related to turmeric	118 (98.33)	2 (1.67)
2	Marketing sources for the turmeric crop	85 (70.83)	35 (29.17)
3	Sources of getting input materials for turmeric crop	115 (95.83)	5 (4.17)
4	Calculate the cost of the turmeric produced	110 (91.67)	10 (8.33)
5	Factors that are more profitable in your turmeric enterprise	114 (95.00)	6 (5.00)

Table 7 indicated that, majority 98.33 per cent of the respondents had knowledge of modern technology related to turmeric followed by 1.67 per cent of respondents had never knowledge. Majority 70.83 per cent of the respondents had knowledge about marketing sources of the turmeric crop and 29.17 per cent of the respondents had never knowledge. Majority 95.83 per cent of the respondents had knowledge about sources of getting input materials and 4.17 per cent of the respondents had never knowledge. Majority 91.67 per cent of the respondents had knowledge about calculate the cost of the turmeric produced and 8.33 per cent of respondents had never knowledge. Majority 95.00 per cent of the respondents had knowledge factors that are more profitable in their turmeric enterprise and 5.00 per cent of the respondents had never knowledge.

Table 8: Distribution of respondents according to their knowledge of the enterprise index

Sl.No.	Category	Frequency	Percentage
1	Low (up to 48.60)	15	12.50
2	Medium (48.61 to 90.06)	98	81.67
3	High (above 90.07)	7	5.83
	Total	120	100
	Mean=69.33		

Table 8 indicated that majority of 81.67 per cent of respondents had medium knowledge of the enterprise, whereas, 12.50 per cent and 5.83 per cent of respondents had low and high respectively knowledge of the enterprise of turmeric growers.These results are in conformity with Jenila (2018).

Managerial Assistance

Table 9: Distribution of respondents according to their managerial assistance

Sl.No.	Statement	Always F (%)	Sometimes F (%)	Never F (%)
1	Credit supervisor's discussion with you about your credit requirements.	0 (0.00)	6 (5.00)	114 (95.00)
2	Officer help you to apply for training courses related to turmeric cultivation	6 (5.00)	13 (10.83)	101 (84.17)
3	Anybody help you get loan from bank.	1 (0.83)	60 (50.00)	59 (49.17)
4	Officer come to you or asked to estimate input requirement for your turmeric enterprise.	4 (3.33)	6 (5.00)	110 (91.67)
5	Officer come to you or asked to see him for you information regarding your turmeric enterprise	4 (3.33)	7 (5.83)	109 (90.84)

Table 9 indicated that, majority 95.00 per cent of the respondents had never credit supervisor's discussion with them about their credit requirements whereas 5.00 per cent of had sometimes credit supervisor's discussion with them about their credit requirements. Majority 84.17 per cent of the respondents had Officer never help them to apply for training courses related to turmeric cultivation followed by 10.83 per cent and 5.00 per cent of respondents had sometimes and regularly officer help them to apply for training courses related to turmeric cultivation respectively. Majority nearly half 50.00 per cent of the respondents had sometimes anybody help them get loan from bank, followed by 49.17 per cent and 0.83 per cent of respondents had anybody help them get loan from bank, respectively. Majority 91.67 per cent of the respondents had never officer come to them or asked to estimate input requirement for their turmeric enterprise followed by 5.00 per cent and 3.33 per cent of the respondents has sometimes and regularly officer come to them or asked to estimate input requirement for your turmeric enterprise. Majority 90.84 per cent of the respondents had officer never come to them or asked to see him for you information regarding your turmeric enterprise followed by 5.83 per cent and 3.33 per cent of the respondents had sometimes and regularly officer come to them or asked to see him for you information regarding your turmeric enterprise.

Table 10: Distribution of respondents according to their managerial assistance index

Sl.No.	Category	Frequency	Percentage
1	Low (equal to 0.00)	44	36.67
2	Medium (0.01 to 30.00)	66	55.00
3	High (above 30.00)	10	8.33
	Total	120	100
	Mean=12.33		

Table 10 indicated that, majority of 55.00 per cent of respondents had medium managerial assistance, whereas 36.67 per cent and 8.33 per cent of respondents had low and high respectively managerial assistance of turmeric growers.

Overall Entrepreneurial Behaviour of Turmeric Growers

The respondents were categorized into low, medium and high levels of entrepreneurial behaviour on the basis calculated index range as shown in table 11.

Table 11: Distribution of respondents according to their overall entrepreneurial behaviour index

Sl.No.	Category	Frequency	Percentage
1	Low (Up to 46.09)	15	12.50
2	Medium (46.10 to 64.12)	88	73.33
3	High (above 64.13)	17	14.17
	Total	120	100
	Mean=55.10		

From the Table 11 revealed that, majority 73.33 per cent of the turmeric growers had medium level of entrepreneurial behaviour, followed by high 14.17 per cent and low 12.50 per cent, respectively. From the above trend possible reason for the majority of the respondents in medium entrepreneurial behaviour might be due to their medium land holding and due to not taking a risk and late adoption of new technology beside medium risk bearing ability.

Conclusion

The findings of the research observed that, majority of the farmers had medium entrepreneurial behaviour, which indicates that the turmeric growers are moving towards progress and for transformation of the responded from the medium to high entrepreneurial behaviour, it need support from the government agencies to develop strategies for the entrepreneurs to maximize related income generation of the turmeric growers through turmeric enterprise. There is need to provide advanced agriculture technologies to the turmeric growers and motivate them to adopt that technologies.

References

Anonymous, 2019. District Agriculture Department, Wardha.

Bindu, N., Y. N. Shivalingaiah and Shwetha, N. V. 2019. Entrepreneurial Behaviour of Flower Growers in Tumkur District of Karnataka State. Int. J .Curr. Microbiol. App. Sci., 8(3): 656-663.

Gartner, W., N. M. Carter and P. D. Renolds 2010. Entrepreneurial behaviour: Firm Organizing process. Handbook of entrepreneurship research, 99- 127.

JenilaStephency, T. 2018. Entrepreneurial behaviour of coconut growers in kanyakumari district. EPRA International Journal of Economic and Business Review, 6(5).

Kadte, A. J., D. S. Perke and P. S. Kale 2018. Economics of turmeric production in Sangli district of Maharashtra, India. International Journal of Current Microbiology and Applied Sciences, 6: 2279-2284.

Kaushalkumar (2012) Entrepreneurial behaviour of pineapple growers. Indian Research journal of extension education, 1:142-145.

Kolgane, B. T., S. R. Suramwad and R. V. Dound. 2018. Study of entrepreneurial behaviour of pomegranate growing farmers in Solapur district of Maharashtra state. Journal of Pharmacognosy and Phytochemistry: SPI, 2956-2958.

Madhu, B. M., 2010. Technological gap in turmeric production practices in Bengaum District. M.Sc. (Agri.) Thesis (Unpub.), University of Agril. Sci. Dharwad.

Patel, P., M. M. Patel, S. K. Badodia and P. Sharma 2014. Entrepreneurial Behaviour of Dairy farmers. Indian Research Journal of Extension Education, 14(2):46-49

Tekale, V. S., D. N. Bhalekar and J. I. Shaikh 2013. Entrepreneurial behaviour of Dairy farmers. Indian Journal of Extension Education, 9:32-36.

20

Dairy Unit
A Better Option for Doubling of Farmer's Income

C. Satapathy and Sasmita Panda

Director and Research Associate, Amity, Bhubaneswar, Odisha

Abstract

The study was conducted in Odisha covering 8 districts with a sample of 1750 milk producers to investigate in to the status, problem and factors contributing for preference of dairy farming as better option to double the income of the farmers. The study revealed that milk producers are middle aged, belong to OBC caste structure, live with medium family size, with farming as major occupation having little or no irrigation facility. Majority of them are educated up to high school standard and have good number of years of experience in dairy farming. The selected samples have 3587 crossbreed cows, 894 desi cows & 179 buffaloes for the purpose of milk production. There is significant gap in between potentiality and actual milk yield observed in field level. The adoptions of dairy technology are considered in terms of nutritious feed, fodder and azolla, cleanness of cow shed & treatment of animals. The milk producers need to be sensitized to pay attention for animal health and keeping constant contact with technical people. The major problem in dairy farming are found to be low procurement price, low coverage of animal insurance, non availability of fodder, high cost of feed and lack of timely treatment of animals. The milk producers are satisfied with dairy farming. The contributing factors for positive perception of dairy as better option to increase farm income are found to be whole family engagement, ready market for product, organised efforts of OMFED, high value by-product & milk as everyday requirement of human beings. To improve dairy farming the suggestions include increase in procurement price of milk, provision of subsidised feed and medicine, financial help to construct cow shed, timely payment of insurance amount, training, exhibition and welfare fund for milk producers.

Keywords: *Milk producers, Dairy farming, Animal insurance, Milk yield,Fed*

Introduction

NITI Policy Paper No.1/20017 reveals details of the concept and time frame as how to double the income of the farmers. The policy paper emphasizes in increase in agriculture productivity, improvement in total factor productivity, diversification towards high value crop, and increase in crop intensity, increasing terms of trade for farmers and shifting cultivators to non-farm and subsidiary activities. The well thought and planned program for doubling the income of the farmers reflecting dream or reality. We have farmers of different types like, small. Marginal and large farmers, resource rich and resource poor farmers, farmers of irrigated and non-irrigated zones, tribal and non-tribal farmers , trained and untrained farmers and progressive and non-progressive farmers. The farmers and their economic structure are quite different from place to place even within district. NSSO data on consumption expenditure survey (2011-12) reveal that more than one-fifth of households with self-employment in agriculture and their principal occupation were having income less the poverty line.

Milk Production Scenario

Milk is primary source of protein and lactoseDairy farming is considered to be a whole environment of reciprocal relations and dependency reproduction and protection of social values. In worldwise dairy cows count up to 264 millions producing 600 million tones of milk every year. The global average for milk production is approximately 2200 litters/cow. Cattle population of world is 1009.69 millions heads (FAO, 2019) and highest milk producing countries are India, USA, China, Pakistan and Brazil. In case of consumption of milk India tops in the world followed by EU, USA and China. The per capita consumption of milk is highest in the country of Belarus followed by Ukraine and New Zealand. India is ranked 7th in the world in this respect. In India total cattle population is 306.70 millions. The state of Odisha constitutes 1.19% of the total cattle population of India. The level of milk production of India is reported to be 198.2 metric million tones. The per capita availability of milk in India is 427 gms/day (2021) and highest being in Punjab followed by Rajasthan. In case of Odisha the per capita milk availability is 143 gms and in milk production the district Cuttack top the list followed by Ganjam and Jagatsinghpur.

Considering situation and environment within which rural households work and reside in Odisha the prospect of diary unit is better than crops to double the income of the farmers.

Objectives

Keeping these considerations in view a survey was conducted in Odisha taking dairy farmers of OMFED units of eight districts to examine the prospect of doubling income with following specific objectives.

1. To understand the living environment of milk producers.
2. To ascertain diffusion of dairy technology and their adoption status.
3. To determine the gap between existing infrastructure and requirements to increase milk production.
4. To determine constraints in dairy farming and suggestions to overcome them.

Review of Literature

Dairy development in developing countries has played a major role in increasing milk production, improving income level in rural areas, generating employment opportunities and improving the nutritional standards of the people, especially for small and marginal farmers. Low and unreliable income from cash crops suggest that alternative farming activities should be developed. This is in spite of indications that there is a potential for dairy development, and dairy can reduce the level of poverty. However, smallholder dairy production is becoming increasingly important and it contributes magnificently to the improvement of the livelihoods of rural people. Higher level of technology adoption is associated with better milk yield and improved dairying has a direct impact on income generation, poverty alleviation and availability of animal protein. Thus, to increase the milk production existing dairy technology should be adopted in the small household dairy farms. Dairying is growing faster in Bangladesh but facing the problems of inefficient management practices and health care, lack of high quality breeds, lack of proper breeding programme to improve the existing dairy cattle resource, high input and low output prices leading to lower productivity (Uddin *et al.* 2010). The dairy technologies encompass the use of crossbred animals, improved feed technology and improved management (Mohamed *et al.* 2004). But cattle farming in rural Bangladesh are constituted mainly from smallholder farming system being managed in traditional ways. The effect of several technical (breeds, artificial insemination, vaccination, etc) and socio-demographic factors would be beneficial to improve the dairy production. Understanding the factors affecting farmers' adoption of dairy technology is critical to success of development and implementation of policies and programmes in dairy industry development. Previous adoption oriented research has examined the use and diffusion of dairy-related technologies (Metz *et al.* 1995, Nicholson *et al.* 1999; Mohi and Bhatti 2006; Mekonnen *et al.* 2009) and the factors affecting adoption on smallholder farms (Irungu *et al.* 1998; Jera and Ajayi. 2008). However, numerous studies have found that constraints But surprisingly little work has been done to examine the status and constraints of adoption of dairy technologies in rural smallholder dairy farms in Bangladesh. The nation still has the potential to meet the growing demand for milk, but the immediate need is to adopt and follow better technologies of dairy farming. To increase the country's milk output, a judicious strategy of focusing on high yielding breeds and improved management technology should be adopted for considerable dairy development. Before that, it is need to know the base-line information of existing features of technology adoption by the small dairy farmers and the related constraints to adopt these technologies. The present study is, therefore, undertaken to identify the causes of adopting and non-adopting crossbred cattle, adoption level, effect of adoption on milk yield, factors that influence in adoption and various constraints encountered by the farmers.

Methodology

Sample of the Study: The members of Dairy Cooperative Societies (DCS) constituted sample of the study. The study was undertaken in eight districts of Odisha covering 150 DCSs out of 298 and as many as 1750 members of DCS participated in interview programs offering their free and frank opinions. The members having a minimum of five years' experience in dairy farming were selected from 150 DCSs as respondents. The sample contains 712 male (40.69%) and 1038 female (59.31%) milk producers.

Result Discussion

1. **Living Environment of Milk Producers:** The selected sample respondents consisted of 1750 milk producers being members of DCSs under the domain of OMFED Odisha. Majority of them are found to be middle aged and belonged to OBC caste categories. Majority of them have education up to middle school and High School standard and very few were college educated. They lead a family of 4-5 members and were from small and marginal farmer group. Their major occupation is farming and few of them have irrigation facilities. The fodder cultivation has not yet been popular among them. As members of DCSs they have been exposed to training on dairy farming and management of cooperative societies.
2. **Animal Status:** The sample 1750 milk producers earn their livelihood from dairy farming. The possession of cows and buffaloes of the sample respondents were found as contained in table.

Table 1: Possession of Cows and Buffaloes

Animal	Milching Stage	Dry Stage	Total
1. Cross Bred	2580	1007	3587
2. Desi cow	671	223	894
3. Buffaloes	68	111	179
Total	3319	1341	4660

The milk producers under study rear C.B cows and desi cows and buffaloes for the purpose of milk production and sell milk to OMFED. The milk production status of the sample animals reveal the gap as mentioned in table.

Table 2: Gap in Milk Production

Animal	Milk Production potential per day per each	Milk Production actual per day per each	Gap (%)
1. C.B. Cow	8.00	4.68	41.50
2. Desi Cow	3.00	1.62	46.00
3. Buffaloes	10.00	7.50	25.00

The milk potential per day per animal was ascertained from the progressive milk producers in focus group discussion under agro-climatic condition in which they reside and rear animals.Even against local potentiality of milk production the actual production level of milk is much less with a considerable gap from 25 to 46 percentage.

3. Diffusion and Adoption of Dairy Technology

The dairy farmers are under study are exposed to different sources of dairy information. The major sources are extension agencies, TV, Radio and literature. The DCSs have definite program to make awareness about dairy technology to the milk producers for adoption and increased milk production.

The dairy technologies under study examined, are feed, fodder, cow shed and treatment. Since the breeding through AI is done by all the milk producers this aspect in the context of adoption was not taken into consideration.

Table 3: Diffusion and Adoption of Dairy Technology

Feed of Animal	Number Adopted	Percentage
1. OMFED supplied feed (compound)	1750	100.00
2. Fodder	120	6.85
3. Azolla	70	4.00

In case of cattle feed, fodders as well as azolla are used in limited scale. The dairy farmers under study hardly grow fodder to feed the animals. Growing of fodder like Napier grass, Cow pea and maize is not much popular in the area under study.

4. Management of Cow Shed

Space, Light, Ventilation, clean drinking water, cemented floor, regular cleaning of sheds were taken into consideration in case of cows. The buffaloes are kept in open space

Table 4: Management of Cow Sheds (N=1750)

Parameters	Recommendation		
	Fully Followed	**Partially Followed**	**Not Followed**
1. Space for animal	870	675	205
2. Ventilation	700	750	300
3. Light	950	650	150
4. Provision of clean drinking water	1100	415	235
5. Regular cleaning	1200	250	300
6. Regular bathing of animals	680	510	560

Management of cow shed appear to be less than satisfactory as fully followed is much less than partially and not at all followed taking together.

5. Treatment of Animals

Cows and buffaloes suffer from a variety of diseases. The common treatment like vaccination, deworming, black quarter, foot and mouth disease, pox, lesions, ring worm, milk fever are common in the area under survey. The degree of alertness and regularity in treating the animals was studied with following results.

Table 5: Treatment of Animals

Degree of Regularity in Treating Animal	Frequency	Percentage
1. Most regular	428	24.46
2. Regular	909	51.94
3. Not regular	215	12.29
4. Not at all regular	198	11.31
Total	1750	100.00

The milk producers have not been sensitive to treat the animals regularly. In the scale of regularity the most regular is only 24.46 percentage and not at all regular is up to 12.29 percentage and that too not at all regular counts up to 11.31 percentage. There is need persuade milk producers to treat the animal in time.

6. Diffusion of Technology

Technology on dairy technology can be diffused successfully if the milk producers visit animal health centres on priority basis looking at the illness of the animals. The frequency of milk producers to veterinary dispensaries/ hospital as the case may be was examined.

Table 6: Visit of Milk Producers to Animal Health Care Centres (N=1750)

Degree of Frequency in Visiting Animal Health Care Centre	Mentions	Percentage
1. Very frequently	112	6.40
2. Frequently	401	22.91
3. Some times	704	40.23
4. Only on emergency case	533	30.46
Total	1750	100.00

Results indicate that visit of milk producers to animal health care centre is mostly some times and on emergency cases. As understood during interview that milk producers also practice traditional method of treatment. They feel local wisdom also can give relief to animals in cases of minor diseases. However, there is need to motivate milk producers to be in contact with Veterinary doctors to reduce risk factor.

7. Problems of Dairy Farming

There are multiple problems in maintaining dairy unit in rural areas. There is no end to listing of problems as expressed by milk producers of different districts owing variation in agro-climatic condition. Most important problems were screened during filed data collection are given in table.

Table 7: Problems in Dairy Farming (N=1750)

Problems	Frequency	Percentage	Rank
1. High Cost of feed	323	18.46	IV
2. Non-availability of fodder	415	23.71	III
3. Contact with veterinary Doctors	284	16.23	VI
4. Timely Treatment	297	16.97	V
5. Maintaining quality of milk	219	12.51	VII
6. Disposal of milk	70	4.00	VIII
7. Low procurement price by OMFED	1650	94.28	I
8. Insurance of animal	897	51.26	II

The problems mentioned in the table are quite valid as understood in focus group discussion. The low procurement price, availing of insurance amount, non-availability of fodder, high cost of feed and contact with veterinary doctors at time of emergency are the important problems faced by the sample respondents

8. Contributing Factors for Perceiving Dairy as Better Option

At present farmers mostly small, marginal and large farmers in rural area have accessibility to number of options to earn livelihood. The reasons for attaching more importance to double the farm income was ascertained as reflected in table below

Table 8: Contributing Factors for Perceiving Dairy as Better Option

Factors	Score Value	Percentage
1. Family engagement	2.51	11.20
2. Income throughout the year	2.62	10.65
3. Ready market	2.77	11.26
4. Every day requirement	2.39	9.71
5. Meeting home consumption	2.16	8.78
6. Organized efforts by OMFED	2.68	10.89
7. Insurance against death of animal	2.34	9.51
8. High value by-product	2.46	10.00
9. Scope for daily/weekly earning	2.38	9.68
10. Religiously and culturally acceptable	2.29	9.32
Total	24.60	100.00

The contributing factors towards perception of dairy as better option for doubling farm income are more or less closed to each other indicating inter dependency of variables ranging from 8.78 to 11.26 percentage. The ready market for milk, whole family engagement, organized by OMFED, income throughout the year, high value by-product,every day requirement of milk, scope for daily/weekly income and compatibility to our social system are the important contributing factors for positive perception.

9. Satisfaction of the Milk Producers

The satisfaction of the respondent about dairy farming was ascertained to know the feeling and attitude towards dairy compared to other livelihood system.

Table 9: Satisfaction of the Respondents

Degree of Satisfaction	Number	Percentage
1. Very much satisfied	270	15.93
2. Satisfied	1128	64.46
3. Neutral	205	11.71
4. Dissatisfied	133	7.60
5. Not at all satisfied	14	0.80
Total	1750	100.00

On the whole,the satisfaction of milk producers under study towards dairy farming is more thandissatisfaction.

Suggestions for Improvement

The respondents offered a number of suggestions to make dairy farming more profitable and sustainable. The suggestions are (i) Increase in procurement price of milk by OMFED (ii) Provision of subsidized feed and medicine (iii) Financial help to construct cow shed(iv) Payment of insurance amount in case of death of animal (v) More of training programs (vi) Local exhibition for competition in dairy units and (vii) Welfare fund for the milk producers.

Conclusions

1. The study 'Dairy unit: a better option for doubling of farmer's income' conducted in eight districts of Odisha with a sample of 1750 milk producers of DCSs lead to arrive at the following conclusions .
2. The women dairy farmers are comparatively more in composition of milk producers. The majority of milk producers are small and marginal farmers and belong to OBC caste structure. At present dairy is the major source of income of the sample respondents.
3. Now in rural areas CB cows are more than desi cows. The number of desi cows is decreasing at a faster rate.
4. There is significant gap in between potential and actual milk yield which need to be bridged.
5. The rate of diffusion and adoption of dairy technologies is relatively low indicating intensification of extension efforts.
6. The major deficit areas in dairy farming are found to be management of cow sheds, timely treatment of animals, and growing of sufficient fodder crops.
7. The milk producers face the problems like low procurement price of milk., uncertainty of insurance amount, high cost of feed, timely treatment of animals and technical support .
8. The contributing factor for positive perception of dairy as better option are found to be the reasons whole family engagement, ready market for product, high value by-product, socio-cultural compatibility and assured income throughout the year
9. The milk producers are comparatively more satisfied with dairy farming and have few suggestions to make it more profitable. These ar3e increasing procurement price of milk, provision of subsidized feed and medicine, assured insurance benefit, financial help to construct cow shed, training and welfare fund.

References

Jera R and Ajayi OC (2008). Logistic modelling of smallholder livestock farmers' adoption of tree-based fodder technology. Agrekon, 47: 379-382.

Kelay B (2002). Analysis of dairy cattle breeding practices in selected areas of Ethiopia. PhD Thesis, Humboldt University of, Department of Animal Breeding in the tropics, Berlin, Germany: 6-108.Khan MJ, Peters KJ.

Khanal AR, Gillespie J and MacDonald J (2010). Adoption of technology, management practices, and production system in US milk production. J. Dairy Sci. 93: 6012-6022.

MM, Sultana MN, Ndambi OA, Hemme T and Peters KJ (2010).A farm economic analysis in different dairy production systems in Bangladesh. Livestock Research for Rural Development.22: 122.

Mohamed AM, Ahmed SE and Yemesrach A (2004). Dairy development in Ethiopia. EPTD discussion paper No. 123, International Food Policy Research Institute, Washington, DC, USA.

Nicholson CF, Thornton PK, Mohammed L, Muinga RW, Mwamachi DM, Elbasha EH, Staal SJ and Thorpe W (1999). Smallholder dairy technology in coastal Kenya: An adoption and impact study. ILRI Impact Assessment Series 5, International Livestock Research Institute, P.O. Box 30709, Nairobi, Kenya.

Nkenye, Schroeder T and Norman D (1997). Factors affecting adoption of improved maize seed and fertiliser in NorhernTanzania. J. Agric. Econ. 48: 1-12. Pal S (2006). Study on dairy livestock feeding and health care practices among dairy.

21

Perception of Household Food Security Through Livestock and Poultry Enterprises

Deepjyoti Roy[1], M.N. Ray[2], A.K. Chaturvedani[3], Jayant Goyal[3] Manju Sahu[1]

[1,3]Department of Veterinary Extension, FVAS, RGSC-Banaras Hindu University, Mirzapur, Uttar Pradesh
[2]Department of Extension Education, CVSc, AAU, Khanapara, Assam

Abstract

Animal husbandry and poultry enterprises play an important role in ensuring household food security as well as economic engagement for farmers. The core concept of food security argues that everyone has physical and economic access to the basic meals they require at all times (FAO, 2002). A study was conducted in the Goalpara district of Assam on the womenfolk belonging to the three tribes Rabha, Garo, Hajong and one ethnic group, i.e. Koch-Rajbongshi. Snowball sampling was employed in the selection of the respondents. A total of 50 respondents were selected from each group to make the sample size 200. The majority (71.00 percent) of respondents believed their engagement ensured household food security. The correlation analysis revealed that social participation ($r= 0.242^{**}$, $P<0.01$), mass media exposure ($r= 0.216^{**}$, $P<0.01$) and attitude toward improved livestock rearing ($r= 0.343^{**}$, $P<0.01$) were highly significant and positive relationship with household food security, while extension contact ($r= 0.162^{*}$, $P<0.05$) had significant and positive relationship. Though the respondents' general opinion of household food security by engaging in livestock and poultry rearing is positive, it is obvious that they are perplexed, and their attitudes toward enhanced animal husbandry in connection to their perception of household food security must be articulated. Government machinery, local authorities, NGOs, and other similar public and private extension agencies can capitalize on the current scenario to alter their current status by changing their attitude and boosting their mass media exposure with proper financing and providing credit facilities.

Keywords: *Animal husbandry, Household food security, Private extension agencies, Mass media, Ethnic group.*

Introduction

Animal husbandry and poultry rearing, in addition to offering economic participation, play an important role in securing household food security. The core notion of food security argues that everyone should have physical and economic access to the basic nutrition they require. 'Food security is a scenario that exists when all people have physical, social, and economic access to sufficient, secure, and nutritious food that fits their dietary needs and food preferences for an active and healthy life', according to the FAO (2002). Consequently, two requirements must be met in order to ensure food security: the first is the availability of food, and the second is the ability to purchase it. There is a propensity to define food security at the national level without examining its condition at the grass roots level with major emphasis at the household or individual level. This is because, while national food security may appear to be adequate, disparities in food security may exist among regions, groups, households, and people.

Increased crop output can increase people's calorific intake, but protein is also essential for health, and pulses and coarse grains are essential. According to the Directorate of Economics and Statistics, DAC & FW, total pulse output in the nation was roughly 25.42 million tonnes in 2017-18, and over 23.40 million tonnes in the following fiscal year (2018-19). This demonstrates the country's falling pulse production pattern, diminishing the availability of vegetable protein to the public. A similar trend was observed in states like Assam, where the area under pulse production declined from 1,54,706 hectares in 2017-18 to 1,50,229 hectares in 2018-19, and further fell to 1,43,962 hectares in 2019-20. The provisional data further forecasted the area to fall even further to 1,42,303 hectares in 2020-21. (Statistical Handbook of Assam, 2020 and 2021)

Food Security Through Livestock and Poultry Enterprises

The management, processing, and sale of livestock products provided more financial gains compared to most of the other tasks conducted by women, which benefited the entire family by boosting food security at the household level (Patel *et al.*, 2016). Begum (2019), in her study reported that the majority of tribal women (91.19 per cent) felt a medium degree of household food security through animal products, whereas 5.71 per cent and 3.41 per cent perceived a high and poor level of food security, respectively. The mean and standard deviation were 38.55 and 6.88, respectively. A study on impact of female participation in livestock and poultry enterprises on women empowerment and household food security was conducted in the Goalpara district of Assam on the womenfolk belonging to the three tribes Rabha, Garo,

Hajong and one ethnic group, i.e. Koch-Rajbongshi. Snowball sampling was employed in the selection of the respondents. A total of 50 respondents were selected from each group to make the sample size 200. The majority (71.00 percent) of respondents believed their engagement ensured household food security. The correlation analysis revealed that social participation ($r= 0.242^{**}$, $P<0.01$), mass media exposure ($r= 0.216^{**}$, $P<0.01$) and attitude toward improved livestock rearing ($r= 0.343^{**}$, $P<0.01$) were highly significant and positive relationship with household food security, while extension contact ($r= 0.162^{*}$, $P<0.05$) had significant and positive relationship. As there were several dimensions of women empowerment in the study, it is one thing to promote women empowerment and quite another to perceive food security. In other words, respondents' responses on household food security were merely evaluated through the lens of a single dimension of protein of animal origin contributing to it. However, because women's empowerment is a multifaceted phenomenon, respondents' perspectives were found to be diverse. As a result, while half of the respondents had a negative or indifferent attitude toward better animal husbandry, their opinion that livestock and poultry enterprises may promote women empowerment and family food security was a positive indicator (Roy, 2022). The predominant focus of prior studies has been on the function of women in agricultural production, with little acknowledgement of their involvement in animal husbandry in comparison to males and children. Women are frequently in charge of supplying basic subsistence needs, whereas males dealt with marketable items. Thus, access to home-produced milk or eggs by mothers is indisputable proof of infant nutrition (Tanka et. al., 2000). Bardhan, *et al.* (2006) identified the factors contributing to food insecurity as poverty, a lack of gainful employment, insufficient land distribution, relatively limited access to suitable technologies and other productive resources, and fluctuations in the availability of food stocks over time in the country under nutrition. The downward trend in food grain production growth rates throughout the 1980s had major implications for national food security. Per capita intake of pulses and grains, a primary source of vegetable protein used to augment cereal protein, has declined from 60.70 gm in 1951 to 26.4 gm in 2007. On the other hand, per capita consumption of many animal food items grew between 1983 and 1990, with the exception of goat meat and mutton, which witnessed a little reduction. A person's protein consumption should be around 1 gram per kg body weight for sufficient nutrition, with animal protein accounting for 30-50 per cent of the dairy protein intake to produce an ideal rate of amino acids. They believe that, in addition to providing nutritional security, livestock plays an important role in maintaining food security by 1) generating employment, 2) supplying draught power, 3) providing plant nutrition, 4) providing fuel and biogas, and 5) assisting in weed management.As per Njuki and Miller (2012), livestock interventions are frequently thought to increase food security through alleviating child nutrition. Women's livestock goods are more likely to be consumed by the family than men's livestock products. Women may not have decision-making authority over the use of products for home consumption or sale if they do not own the livestock.According to Kumar (2013), women produce 60-80 per cent of all food in underdeveloped nations. In addition to agriculture, they are interested in associated industries such as dairy, piggery, poultry, fisheries, sericulture, and pisciculture. Rural women are the most significant productive labour force in the economies of the majority of developing countries, including India. Women account for 48.5 per cent of the overall population but constitute 70 per cent of the labour force in animal husbandry. The role of women as custodians of traditional knowledge about seeds is admirable. However, rural women confront several challenges in carrying out these duties. First, women are not given adequate acknowledgment for the aforementioned difficulties. Women possess less than 2 per cent of the land in the country. According to available data, just 5 per cent of extension programmes are directed towards rural women. Furthermore, the majority of extension services are centred on cash crops rather than food and subsistence crops, which are the primary concern of women farmers and the key to food security.The extent to which livestock contributes to food security is dependent on intra-household dynamics such as:

1) Women's ownership of assets, including livestock, and the extent to which they participate in decision making regarding the use of assets, their products, and the income derived from these assets.
2) Women's decision-making on how much stuff to sell and how much to keep for household consumption. They also stated that strengthening women's control over assets (chiefly land and money) had a favourable impact on food security, child nutrition, and education, as well as women's personal well-being (Juliet *et al.*, 2015).

Conclusion

It is visible that the farm women are perplexed, and their attitudes toward improved animal husbandry in relation to their perception of household food security need to be articulated. The current situation can be capitalised by the government machinery, local authorities, non-governmental organizations, and other such public and private extension agencies for transforming their current status by changing their attitude and increasing their mass media exposure with adequate resources. However, for such programmes to be successful, general education, particularly among women, must increase. These steps will undoubtedly let them realize how cattle and poultry enterprises help them feel more empowered and secure in terms of home food.

References

Bardhan, D.; Nath, S.; Dabas, Y.P.S. and Kumar, A. (2006). Livestock International. A Monthly Magazine for Global Perspective on Livestock Health and Population, Vol. 10, No. 1, Pp. 14-23.

Begum, G. (2019). Livelihood sustainability of rural women through livestock enterprises - an analytical study in Goalpara district of Assam. Ph.D.

DAC & FW (2019). Agricultural Statistics At A Glance, 2019. Directorate of Economics and Statistics, Department of Agriculture, Cooperation and Farmers Welfare, Ministry of Agriculture and Farmers Welfare, GOI.

FAO (2002). The State of Food Insecurity in the World 2001. Rome. Retrieved 27th December 2021 from the site: https://www.fao.org/3/y4671e/y4671e06.htm#fn31.

Juliet, K.; Jemimah, N.; Samuel, M. and Elizabeth, W. (2015). Women Livestock ownership and Food Security. Excerpt from the book Bridging the Gender Gap: Women, Livestock Ownership and Markets in Eastern and Southern Africa Produce by ILRI from internet.

Kumar, P. (2013). Empower women for food security. Kurukshetra, Monthly Journal Ministry of Rural Development, Vol. 61, No. 10, Pp. 28-30.

Njuki, J. and Meller, B. (2012). Livestock and Gender: Achieving Poverty Alleviation and Food Security through Livestock Policies that benefit women. Commissioned by GALVmed, Royal Veterinary College, London.

Patel, S. J.; Patel, M.D.; Patel, J.H.; Patel, A.S. and Gelani, R.N. (2016). Role of women gender in livestock sector: A review. Journal of Livestock Science, 7: 92-96.

Roy, D. (2022). Impact of female participation in livestock and poultry enterprises in ensuring Women empowerment and Household food security among selected tribes/ ethnic group in Goalpara district of Assam. MVSc.

Statistical Handbook Assam (2020). Statistical Handbook Assam 2020, Directorate of Economics and Statistics, Govt. of Assam, Beltola, Guwahati.

Statistical Handbook Assam (2021). Statistical Handbook Assam 2021, Directorate of Economics and Statistics, Govt. of Assam, Beltola, Guwahati.

Tanka, F.K.; Jobbar, M.A. and Shapiro, B.I. (2000). Gender roles and child nutrition in livestock production system in developing countries; A critical review. Socio-economic and policy Research working paper, No. 27, Nairobi; ILRI.

22

Study on Milk Marketing Model of East Khasi Hills District Cooperative Milk Union Limited in Meghalaya

Binodini Sethi, Pavan Kumar Jat, Ram Singh and N. Anada Kumar Singh

Agriculture Economics, School of Social Sciences CPGS-AS, CAU, Umiam Meghalaya

Abstract

India is the world's largest milk producer and one of the most promising and exp and ingmarkets for milk and milk products. Indiaranked 1st in milk production contributing 23 per cent of global milk production. Indian dairy cooperatives society have in stalled milk processing of 66.3 millionliters/day, privatesectorenterprisesof 73.3 million liters/day and producer companies of 2.5 million liters/day. In India, totalmilk production is recorded to be 209.96 million tonnes during the year 2020-21. Theper capita availability of milk is accounted to be 425 gms/day in India during the year2020-21. Global milk production was found to be 906 million tonnes during the year2020. World per capita availability of milk is 315 grams / day. In India, around 61 percentof the total milk production is marketed by sweet shops, dudhiyas (middleman), andprivate dairies. Merely 21 per cent of the milk is marketed through the cooperativesocieties. Keeping in view, following objectives were proposed viz., (1) Study on milkprocurement and distribution model in East Khasi Hills District Cooperative Milk UnionLimited (2) Constraints faced by different stakeholders during the procurement anddistribution of milk. The state of Meghalaya was purposively selected out of which twodistricts namely East Khasi Hills and Ri-Bhoi were selected for data collection purpose.The Na Riorganization and their out lets were in terviewedduringprimarydatacollection. The overall performance of the East Khasi Hills District Co-operative MilkUnion Limited was found to be 3,41,55,600 per annum in terms of net profits. Theorganizational structure of the East Khasi Hills District Co-operative Milk Union Limitedwas found to be well set with proper allocation of duties and responsibilities for efficientmanagement. In the study area, three marketing channels were identified in which thechannel-IIIwas more effectivewith the marketingefficiency of44.83 as compared tothe other channels. Onotherh and, maximum quantity of milk was distributed through channel-Icontributed 87.70 per cent of total milk distribution. Based upon the conclusions of the study it was suggested that to in crease the economicstability of the East Khasi Hills District Cooperative Milk Union Limited (NaRi) throu ghprocurement and distribution of milk from it soperational areas, minimization of procurement and distributioncos to fmilk shouldbetaken.

1. Introduction

Milk is the principal source of protein and lactose in the diet. Dairy farming is viewed as a whole ecosystem of reciprocal relationships and reliance, as well as there production and safe guarding of societal values. Dairy farms are primarily made up of high-yielding dairy cows. Goats, sheep, and camels are some of the other animals employed in commercial dairy production. Milk has always been regarded as a healthy food. Milk includes more than9 other essential elements, including protein, carbs,vitamins, minerals, and lipids, in addition to calcium, which makes it critical for your bones.

Every year, about 600 million tonnes of milk are produced by over 264million dairy cows across the world. The global average for milk output per cow is around 2,200 litres. India is the world's top milk producer followed by the United States, China,Pakistan, and Brazil. The economy of India is mostly dependent on agriculture and animal husbandry, with milk production playing a significant rolc. It is also the world's largest producer of buffalo milk (FAO, 2019).

Co-operative dairy sector proved to be resilient as it continued to procuremilk from farmers and supply it to the consumers despite several logistical and other challenges. Milk production in India did not show any sign of slowdown. Milk productionis anticipated to be around 209.96 million tonnes in 2020-21, an increase of about 6.3 percent over the previous year. The estimated per capita availability of milk is expected to increase to about 425 gms /day which is more than the world average of around 315gms/day(NDDB,2020).The dairy co-operatives had a difficult year due to an influx of more milk production, a reduction in milk sales, and the building of inventories of preserved goods. Throughout the year, milk procurement by dairy co-operatives increased as it accepted diverted extra milk from farmers that would otherwise be sold to private and unorganized businesses. During the 2020-21 budget year, milk procurement by dairy co-operativesincreased by 7.9 per cent. Indian dairy co-operatives have installed milk processing capacity of 66.3million litters per day,73.3million liters/day in private sector enterprises, and 2.5 million litres per day in producer companies.

The unorganized milk industry, these businesses have extensive procurement and distribution capabilities. Milk collecting and chilling units at the village level are part of the distribution networks. Milk was gathered at pasteurization, standardization, branding, and packing are all steps inthe processing of these centres at dairy facilities and the creation of some high-value items an estimated 48 percent of India's total milk output is exported. The remaining 52 per cent was either eaten at the producer level or sold to tiny shops in rural areas destined for processing before being sold to urban customers. India is the world's largest milk producer and one of the most promising and expanding markets for milk and milk-related goods. With the yearly growth in per capita income,the demand for milk products has significantly grown. However,compared to the size of her domestic market and milk output, India's contribution toglobal dairy commerce is insufficient. One of the main barriers to the growth and diversification of dairy commerce on the global market is the poor productivity of milch animals, as assessed in India by the amount of milk produced per day per animal. Due to increased imports of dairy products at higher prices and volumes, it also affects the dairy industry's ability to compete globally (Kumar, 2020).

Average milk production in a dairy home was roughly 36 liters per day,which was much more than the national average of 10 liters per day per animal. 207gms/day of liquid milk were used daily per person in producer families, which was more than was made available to each person in the area. Of the 10 per cent of milk output that was still eaten at the home level, 40 per cent was consumed as liquid milk and 60 percent was transformed into milk products. The smaller herd size category serving rural families without a dairy farm had a greater representation of the latter group. The principal marketing channels, which accounted for 61 per cent of the total excess that was marketed, were halwas, dudhiyas, and private dairies. Only 21 per cent of the milk was sold through cooperatives.Despitethefact that the marketed surplus per household was significantly larger (24.02 liter/day), the percentage of milk sold by small farmers through the latter channel was still lower per cent. Dairy co-operative membership had the most impact on small farms' marketed surplus, according to a review of the factors (Kumari*etal.,* 2021).

Milk production and percapita availability of milk in India

Year	Production (MillionTonnes)	Per Capita availability (gms/day)
2015-16	155.5	337
2016-17	165.4	355
2017-18	176.3	375
2018-19	187.7	394
2019-20	198.4	406
2020-21	209.96	425

Source:(NDDB,2020)

2. Methodology

The present study was conducted in the state of Meghalaya. Shillong is the capital of Meghalaya which is located at an altitude of 1,496 meters above sea level. The present study has therefore been purposively selected forEast Khasi Hills District Co-operative Milk Union Limited (NaRi),a popular dairy in the East Khasi Hills district's Mylliem block in Meghalaya. The dairy unit was established in the year 2009 and it started functioning as a business module on 1st September 2017. A study of this primary dairy unit in relation to procurement and distribution has not yet been done by any other researcher, and therefore it was thought appropriate to choose this dairy unit for the present study.

Marketing Cost

The overall cost paid by the producer seller and the many middlemen engaged in the sale and purchase of commodities until the commodities reached the end consumer was calculated as

$$C = C_F + C_{m1} + C_{m2} + C_{m3} + \dots + C_{mi}$$

$$C = C_F + \sum C_{mi}$$

Where,

C = Total cost of marketing of the commodity

C_F = Cost paid by the producer at the time the produces leave the farm till he sold it

C_{mi} = Cost incurred by the i^{th}middleman in the process of buying and selling the product

Marketing Margin

It is the different between the total payments (cost + purchase price) and receipts (sale price) of the middlemen (ith agency). It is expressed as:

Ami = Pri – (Ppi + Cmi)

Where,

Ami = Absolute marketing margin of ith middlemen

Pri = Total value of receipts per unit (sale price)

Ppi = Purchased value per unit (purchased price)

Cmi = Cost incurred on marketing per unit

Price Spread

The difference between the price paid by consumers and the price received by the producer for a comparable amount of agricultural produce, is determined using the sum of average gross margin method.

It is the difference between the price paid by consumer and the price received by the producer for an equivalent quantity of farm product. It is calculated by using the following formula.

Ps = (Pr – Pf)/PrX100

Where,

Ps = Producer's share in the consumer rupee

Pf = Net price receive by producer per unit of output

Pr = Retail price per unit of output

Pf = PA – CF

Where,

Pf = Net price receive by producer

PA = Selling price of the producer

CF = Marketing cost incurred by producer

Price spread = Pc – Pf

Where,

Pc = price paid by consumer

Pf = Net price receive by producer

Sum-of-Average Gross Margin Method

The overall marketing margin and the breakup of the customers' rupee will be calculated by adding the average gross margins of all intermediaries.

$$MT = \sum_{i=1}^{n} \frac{Si - Pi}{Qi}$$

Where,

MT = Total Marketing Margin

Si = Sale value of a product for ith intermediary

Pi = Purchase value to be paid by the ith intermediary

Qi = Quantity of the product to be handled by the ith intermediary

i = 1, 2, 3 … N (Number of intermediaries involve)

Marketing Efficiency

Marketing efficiency is a metric for measuring market performance. The Shepherd's Formula and the Acharya and Agarwal index were used to assess marketing efficiency.

a) Shepherd's Formula

According to the equations, the ratio of the total value of goods marketed to the marketing cost may be used to determine marketing efficiency. The greater this ratio, the greater the efficiency, and vice versa.

This can be expressed in the following form:

ME = [V/I] - 1

Where,

ME = Index of marketing efficiency

V = Value of goods sold

I = Total marketing cost

b) Acharya and Agarwal index

The following formula can be used to compare the relative efficiency of different markets:

E = O / I

Where,

E = marketing efficiency expressed in percentage terms by multiplying with 100.

O = value added to outputs in the marketing system and

I = inputs used in the marketing process

$$\text{Depreciation(₹/year) x} = \frac{\textit{Original value of an asset} - \textit{Junk value}}{\textit{Useful life of an asset}}$$

3. Result

The milk was collected by East Khasi Hills District Cooperative Milk Union Limited (NaRi) from 21 dairy co-operative societies (DCS) from different districts of Meghalaya. It was found that the highest amount of milk procured from East Khasi Hills District DCS followed by Ri-Bhoi district DCS, West Garo Hills district DCS and Jaintia Hills District DCS respectively. NaRiorganisation was procuring 13,500 litres of milk per day from various DCS from four districts. Out of total milk procured by NaRi (90%) of milk were sold out through NaRi outlets and (10%) milk was utilisedfor value-added products per day. NaRi organization was found to purchase raw milk from DCS at the rate of ₹ 47 per /litre and registered outlets of NaRi was purchasing milk from NaRiorganisation at the rate of ₹ 54.50/litres. The outlets of NaRiorganisation were selling milk packets to consumers at the rate of ₹ 60/litres.

Utilization Pattern of Procured Milk of NaRi

The NaRiorganisation working on procured milk from dairy co-operative society (DCS). The daily procurement of milk was recorded to be of 13,500 litres, whereas, monthly procurement of milk was calculated to be 4,18,500 litres. From the table . it was observed that 1,350 litres of milk procured was utilized for value addition by NaRi and 12,150 litres were pasteurized milk per day which were used to sell through registered outlets of NaRi.

From this it can be interpreted that out of the total milk procured more than 90 per cent were pasteurized and the remaining 10 per cent was utilized to make various value-added products like courd, paneer, cream *etc.,* which can provide a better economic advantage in terms of profit to the NaRi organization.

Utilization pattern of procured milk of NaRi

Sl.No	NaRi Organization	Litres/ day	Litres/ month
1.	Total milk procured by Na Ri	13,500	4,18,500
2.	Pasteurized milk for selling	12,150	3,76,650
3.	Milk used for making value added products	1,350	41,850

(Source: Primary data collected from NaRi Organization).

Marketing pattern, Procurement and Distribution of Milk in Study Area

During the study three channels were identified based on the distribution pattern of milk to NaRi. These channels were as follows,

1. **Channel – I**: Dairy Co-operative Society → NaRi Organization → Outlets → Consumer
2. **Channel – II:** Dairy Co-operative Society → Halwai
3. **Channel – III:** Dairy Co-operative Society → Consumer

From this it is understood that the dairy co-operative society was the starting point from where the milk was started to distribute. Dairy co-operative societies obtain the milk from its farmers and from there it was distributed further. So, in the case of all the channels DCS was involved. Channel-I was with 4 functionaries which comprised of DCS, NaRi, Outlets of NaRi and finally consumers. Channel-II which included the DCS and halwai those who collected milk from DCS without involvement of NaRi Organization. Channel-III was the only channel through which consumers bought milk directly from DCS without any intermediaries.

Marketing Cost Incurred by Dairy Co-operative Society

The results of study on marketing cost incurred by dairy co-operative society which was in all the three channels was elucidated in the Table. The marketing cost incurred per litre of milk showed that cost was incurred as transportation and labour charges. ₹ 4.53, ₹ 3.56, ₹ 1.2 were the respective marketing cost by DCS n corresponding channels. DCS had to spend ₹ 3.53 in in channel-I and ₹ 2.36 in channel- II per litre of milk for transportation which incurred the major share in marketing whereas in channel-III there is no expenditure from DCS side for transportation since it was sold directly to consumers.

It was interestingly found that DCS received a higher net price per litre (₹ 53.8) of milk sold in channel-III compared to channel-I and II from which they received ₹ 42.47 and ₹ 46.44 respectively. This anomaly is due to the fact that the selling price of milk in channel-I was less (₹ 47) while selling to NaRi, since the DCS has been deriving various benefits due to NaRi.

Marketing Cost Incurred by Dairy Co-operative Society
(Value in ₹/litre)

Sl.No	Particulars	Channel-I	Channel-II	Channel-III
		Dairy Co-operative Society		
1.	Transportation	3.53	2.36	0.00
2.	Labour	1.00	1.20	1.20
3.	Marketing cost incurred by DCS	4.53	3.56	1.20
4.	Selling price of DCS	47.00	50.00	55.00
5.	Net price received by DCS	42.47	46.44	53.8

(Source: Author's computation)

Marketing cost Incurred by NaRi Organization and Market Margin

The results of study on marketing cost incurred by NaRi was displayed in the Table. The marketing cost incurred per litre of milk for NaRi organization showed that cost was incurred for transportation additionally NaRi has to bear the cost for processing and packaging since they are manufacturing value added products from milk. Channel-I was the only channel that NaRi was involving in which they had to spend ₹ 47 per litre for purchasing raw milk and ₹ 0.7 and ₹ 0.2 for processing and packaging respectively per litre, whereas in channel-II, channel-III there is no expenditure from NaRi side since it was not involved in those channels. So, the total marketing cost incurred per litre of milk by NaRi in channel-I was found to be of ₹ 2.15.

It was found that NaRi was getting a marketing margin of ₹ 5.35 per litre of milk sold, since they are selling pasteurized milk to outlets @ ₹ 54.50 per litre. After purchasing raw milk from DCS @ ₹ 47.00 and at a margin of ₹ 7.50 they were selling to outlets. They were pasteurizing and packaging the milk which caused a marketing cost of ₹ 2.15 and hence the net margin was found to be of ₹ 5.35.

Marketing Cost Incurred by NaRi Organization and Market Margin

Sl.No	Particulars	Channel-I	Channel-II	Channel-III
		NaRi Organization		
1.	Purchasing price	47.00	0.00	0.00
2.	Processing charge	0.70	0.00	0.00
3.	Packaging charge	0.20	0.00	0.00
4.	Transportation charge	1.25	0.00	0.00
5.	Marketing cost incurred by NaRi	2.15	0.00	0.00
6.	Selling price of NaRi	54.50	0.00	0.00
7.	Marketing margin received by NaRi	5.35	0.00	0.00

(Source: Author's computation)

The Table describes the distribution of quantity and share of total milk procured per month in various marketing channels. Out of the total 5,17,206.8 litres of milk procured by DCS major share belongs to channel-I *i.e.,* 86.70 per cent of total milk procured by DCS was sold to NaRi for further processes. Similarly, 8.3 per cent through channel-II in which halwai were included and 5 per cent was to channel-III which was sold directly to consumers.

Quantity and Share of Milk in Different Marketing Channel

	Dairy Co-operative Society			
	Channel-I	Channel-II	Channel-III	Total
Quantity (litres)	4,48,418.29	42,928.16	25,860.34	5,17,206.80
Per cent (%)	86.70	8.30	5.00	100.00

(Value in litres/month)

(Source: Author's computation)

Marketing Cost Incurred by Outlets and Marketing Margin

After pasteurization the packaged milk was transported to outlets for making it available to consumers. Here in Table showed the marketing cost and margin incurred by the outlets in this process. From the table it was revealed that outlets were paying ₹ 54.50 per litre of milk and they were selling it at a price of ₹ 60. In between they had to bear a marketing cost of ₹ 3.20 which included ₹2.20 as storage charge for refrigeration an ₹ 1.00 as labour charge in outlets. After deducting this cost their net marketing margin was found to be of ₹ 2.30 per litre. In the case of channel-II and channel -III the NaRi outlets had no role hence no cost was incurred by outlets in those channels.

Marketing Cost Incurred by Outlets and Marketing Margin

(Value in INR/litre)

Particulars	**Channel -I**	**Channel-II**	**Channel-III**
Purchase price	54.50	0.00	0.00
Storage charge	2.20	0.00	0.00
Labour charge	1.00	0.00	0.00
Marketing cost incurred by outlet	3.20	0.00	0.00
Selling price	60.00	0.00	0.00
Marketing margin received by outlets	2.30	0.00	0.00

(Source: Author's computation)

Price Spread and Marketing Efficiency of Milk

An attempt has been done to calculate the price spread and marketing efficiency which is elucidated in the. It was revealed that in the case of channel-I the consumer's purchasing price was INR 60.00 which highest amongst all the channels. It was due to the fact that through channel-I consumers are getting pasteurized milk with attractive and safe to handle packaging which also provides enhanced storage life. So total marketing cost also found to be in channel-I and was of ₹ 9.88 followed by channel-II (₹ 3.56) and channel-III (₹ 1.20). Because of advanced processing and all, calculation of price spread seen to be having higher value in channel I which was ₹ 17.53.

Calculation of marketing efficiency by acharya and Agarwal method showed that channel-III as the efficient channel since no intermediaries were involved there with an efficiency of 44.83. Finally, the estimation of producer shares in consumer rupees showed producer is getting more benefitted from channel-III where 97.82 per cent of what consumer is spending goes directly to consumer due to absence of intermediaries there. This also establishes the fact that channel-III as the efficient channel which was underlined by less price spread and increased share of producer share in consumer rupees.

Price spread and Marketing Efficiency of Milk

Particulars	Channel-I	Channel-II	Channel-III
Consumers Purchasing Price	60.00	50.00	55.00
Total Marketing Cost	9.88	3.56	1.20
Total Marketing Margin	7.65	0.00	0.00
Price spread	17.53	3.56	1.20
Marketing Efficiency (Acharya Agrawal)	2.42	13.04	44.83
Producer share in consumers rupees	70.78	92.88	97.82

Source: Author's computation)

Total cost and profit of NaRi Organization

Extent of profit and cost incurred by NaRi organisation presented in Table. The total variable cost including milk, cost of labour, transportation cost, fuel charge, raw material, stationary cost and wastage milk was to be INR 1,75,27.602 per month contributed the overall 99.93 per cent of total cost. On other hand, the fixed cost including machinery and building (depreciation) was to be INR 11,691.66 per month contributed merely 0.07 per cent of total cost. Overall, total cost of production found to be INR 1,75,39,294 per month and gross return INR 2,038,5,594 per month generated INR 28,46,300 per month of profit. The Nari organisation was getting significant profit by processing the milk in Meghalaya.

Total cost and profit of NaRi Organization

Sl.No.	Costs	Particulars Value (INR/ month)	Per cent (%)
1.	***Variable cost***		
A	Milk	1,73,68,542	99.03
B	Cost of labour	1,15000	0.66
C	Transportation charge	13,500	0.08
D	Fuel charge	11,310	0.06
E	Raw material	8,500	0.05
F	Stationary cost	3,700	0.02
G	Wastage milk	7,050	0.04
	Total Variable Cost	1,75,27,602	99.93
2.	***Fixed cost***		
A	Machinery	9,625	0.05
B	Building	2,066.66	0.01
	Total Fixed Cost	11,691.66	0.07
3.	Total cost (1+2)	1,75,39,294	100
4.	Gross return	2,038,5,594	
5.	Profit	28,46,300	

(*Source*: Author's computation)

Costs and Returns of NaRi outlets in Study Area

NaRi outlets was selected from different two district, first one East Khasi Hills district and second one Ri-Bhoi district in Meghalaya. five outlets were selected from East Khasi Hills and one outlet selected from Ri-Bhoi district. Total number of 16 registered outlets working under NaRi organisation out of which 6 outlets were selected for study purpose based on maximum amount of milk transaction. Among 6 outlets the Mawiong, Shillong had transacted maximum quantity and highest profit share of milk followed by MawlaiMawdatbaki, Shillong, Umsning, Ri-Bhoi district, Garikhana, Shillong *etc.* Overall total average of marketing cost was ₹ 18,675 selected outlets of NaRi. Average of selling quantity liters/month, gross return and profit to be followed by ₹ 8,825.00, ₹ 5,29,500.00 and ₹ 29862.50.

4. Conclusions

While assessing the milk marketing model of an organization, the most important aspect of judgement will be the performance and economic stability of them. The NaRi organization's performance has found to be praiseworthy since it has been operating in profit with a well-established procurement network. The overall performance of the East Khasi Hills District Co-operative Milk Union Limited was economical in view of its net profits to the extent of ₹ 3,41,55,600 per/year. The organizational structure of the East Khasi Hills District Co-operative Milk Union Limited was found to be well set with proper allocation of duties and responsibilities for efficient management. In additiondifferent areas of management such as resources, finance, and personnel management aspects involved in procurement and distribution of milk were attended efficiently. Transportation was found to be the major constraint which act as a hindrance in the smooth procurement and distribution. There is a matter of concern to reach out to the reasons for not expanding this to other districts. This study on milk marketing model of East Khasi Hills District Cooperative Milk Union Limited in Meghalaya paves the way for further improvements which should emphasize on modern technologies and other efficient marketing strategies for economic profitability of such organizations.

References

Anonymous. (1995). Dairying in India: An Assessment, Quarterly Economic Report of Indian Institute of Public Opinion. 38(3):31-36.

APEDA. (2021). https://www.apeda.gov.in/apedasite/SubHead_Products/Dairy_Products.html. Accessed 02 March 2022.

Bansal, Sonal. (2010). A comparative analysis of milk procurement system followed by cooperative and private sector milk processing units.Thesis, Master of Business Administration (Agricultural Economics), Punjab Agricultural University, Ludhiana,18-44p.

Bhagya, K.P. and Phanindra, S.T. (2015). A study on procurement of milk by Balaji Dairy in Chittoor districtofAndhraPradesh.Thesis,MasterofScience(AgribusinessManagement), Acharya N.G. Ranga Agricultural University, Rajendra nagar,Hyderabad,22-52p.

Bhanja, S.K. and Dubey, V.K. (1983). Socio-economic problems in milk procurement,Dairy guide,3:27-30.

Callaghan, S., Connor, D. O. and Goulding, D. (2022). Raw milk collection in Ireland: insights into the challenges facing the industry. British Food Journal.

Chahal, S.S. and Gill, K.S. (1993). Price spread in milk marketing in co-operative and pricesector in Punjab.Indian J.Agril.Mktg.7(1):57-62.

Chauhan, A.K., Arora, V.K. and Singh, K. (1991). Economic analysis of various milk procurement systems- Acomparative study. Indian J. Agril. Res., 25(1):29-35.

Dagdeviren, O. G. (2016). Determination of production cost of cow milk and functional analysis of factors affecting milk production in the Bafra district. J. Agric. Sci.,21(2).

Dairy and products Annual (2021). https://apps.fas.usda.gov/newgainapi/api/Repot Dairy and Products Annual_New Delhi_ India_10-15-2021.pdf

Dairy Development in North East India (2018). https://www.vethelplineindia.co.in/dairy-development-in-north-east-india/.Accessed22 Nov,2021.

23

Challenges to Sustaining Beekeeping Livelihoods in Kumaon Hills of Uttarakhand

Ishita Mishra[1], Kiran Rana[2] and Harshit Palival[3]

[1]*Department of Agricultural Communication, Sardar Vallabh Bhai Patel University of Agriculture and Technology, Meerut, Uttar Pradesh*
[2]*Department of Agricultural Communication, Govind Ballabh Pant University of Agriculture and Technology, Pantnagar, Uttarakhand*
[3]*Department of Agriculture Extension Education, Banaras Hindu University Varanasi, Uttar Pradesh*

Abstract

The study was conducted to ascertainChallenges to Sustaining Beekeeping Livelihoods inKumaon Hills of Uttarakhand. Nainital district of Uttarakhand was purposively selected for the study as it accounts for the maximum number of beekeepers among all the districts in the Kumaon Hills. Nainital district has eight blocks, out of which Bhimtal block was selected purposively for the study as it has maximum honey production. Top four honey producing villages Jeoli, Gaga, Bhaluti and Chopra were selected purposively from the Bhimtal block. A total 76 respondents were selected by censes method from the four villages having a minimum of five year of experience and minimum number of ten boxes. The study was conducted to identify the constraints faced by the Beekeepers in Kumaon hills of Uttarakhand. From the study conducted it was revealed that slightly more than half of the respondent (68.42%) had medium extent of adoption in terms of recommended beekeeping practices. The major constraints faced by the beekeepers were lack of necessary skills to adopt recommended and lack of trust that the modern recommended practices are useful.

Keywords: Bee, Beekeeping, Honey production, Adoption, Hills

Introduction

Beekeeping is an environment friendly and agro-forestry based occupation that has established itself as an income generating activity and a commercial enterprise which not only promote agricultural and horticultural development but also contributes in upliftment of the rural economy of the country. Apiculture provides enormous potential for income generation, poverty alleviation and sustainable use of forest resources. Traditional beekeeping with *Apisceranais* widely practised by rural people in Uttarakhand who have inherited the tradition from their forefathers and maintain it till this day. The people in Uttarakhand have indigenous knowledge of beekeeping which is passed from one generation to another. In the hill area traditional hives are more suitable than modern hives. Modern beekeeping is infancy in the area, so people need to be encouraged to use modern bee hives for better management of colonies

Uttarakhand is a land of variety of flora and fauna. Beekeeping forms an integral part of the small holder farming system and plays a significant role as a source of additional cash income in subsistence farming. There are around 2,50,000 beekeeping units in India out of which only 8,700 are in Uttarakhand accounting for about 2500 MT of honey production in 2016-17. Uttarakhand is extremely rich in Bee forage plants but the use of this rich resource is not being made properly (Tiwari, *et al.* 2010). Around 20.00 per cent of the Beekeepers in Uttarakhand do not use any medicine for the management of pest and diseases showing that the farmers are not much aware of the advance methods of beekeeping (Khan *et al.*, 2007). Uttarakhand has huge potential for beekeeping due to plenty of flora and suitability of climate. Beekeeping is an ancient practice that is managed traditionally in Kumaon hills of Uttarakhand. Due to the traditional production system the productivity per hive is low in Uttarakhand. Therefore, it is Important to know about the constraintsfaced by the Beekeepers in practicing apiculture in Kumaon Hills of Uttarakhand.

Methodology

The study was conducted in the Nainital district in Kumaon Hills of Uttarakhand. Bhimtal block was selected Purposively for the study as it accounts for the maximum number of beekeepers as reported by the State Beekeeping Research Centre, Jeolikote. Four villages were purposively selected for the study as the villages accounted for the maximum Honey production in the district. The respondents were selected on the basis of census of beekeepers carried out by State Beekeeping Research Centre, Jeolikote. (2016). A list of registered beekeepers of the selected districts was obtained and accordingly the number of beekeepers were selected by census method from each village. A total sample of 76 Beekeepers was obtained from all the four villages which included 40 respondents from Jeoli village, 17 respondents from Chopra, 14 from Bhaluti and 5 from Gaga village. The selected beekeepers were interviewed regarding the constraints faced by the Beekeepers in various beekeeping practices as specified by National Bee Board with the help of a pre-structured interview schedule. The constraints were enumerated and the frequency score for each practice was recorded and, further converted into percentage.

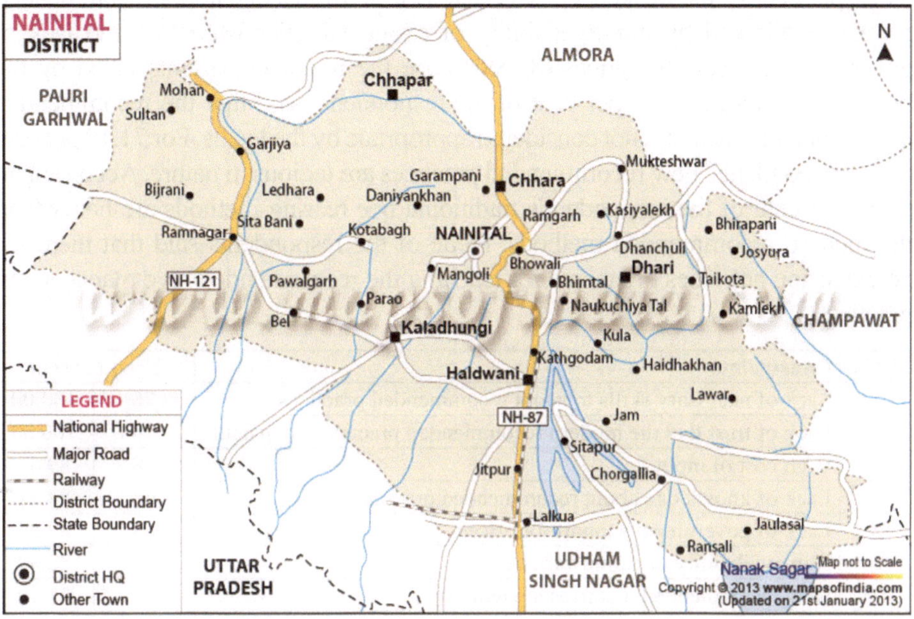

Fig. 1: Map showing Nainital district of Uttarakhand

Result and Discussion

There is a vast potential for beekeeping in the country as well as in the state of Uttarakhand. However, due to lack of knowledge, modern beekeeping practices are not being followed by the Beekeepers. A Beekeeper should always use recommended methods to control swarming, division of colonies, uniting of colonies, mass queen rearing, etc. With the advent of modern recommended beekeeping practices, a great emphasis is being paid by both scientists and extension workers to enhance honey production. Various experiments and researchers have been conducted for gaining high honey production. The resourceful use of recommended technologies to their preference will enable the potential uses i.e the Beekeepers to enhance the honey production, which in turn can improve their economic status. This seems possible only when the beekeepers will adopt the recommended improved cultivation practices according to their preferences.

From the data presented in the Table (1.00) it was found that 100 per cent of the respondents stated that lack of necessary skills to adopt recommended practices and lack of trust that the modern recommended practices are useful are the major constraints faced by the beekeepers in adoption of the recommended practices. High cost of input was another major constraint faced by the beekeepers (89.47%). About 84.21 per cent of the beekeepers said that lack of knowledge about recommended practices followed by non-availability of exotic varieties (82.89%) and lack of government support for prices (82.89%) are the major constraints faced by the beekeepers. About 78.94 per cent of the respondents said that the recommended beekeeping practices are not considered appropriate by the locals. For 71.10 per cent of the respondents' new recommended practices are tedious in nature. According to 67.11 per cent of the respondent's traditional bee rearing methods are better than the modern recommended methods. None of the respondents said that there are no extension activities organized for teaching the recommended hive management practices.

S.No	Constraints	Freq	Percent
1	Lack of necessary skills to adopt recommended practices	76	100.00
2	Lack of trust that the modern recommended practices are useful	76	100.00
3	High cost of input	68	89.47
4	Lack of knowledge about recommended practices	64	84.21
5	Lack of government support for prices	63	82.89
6	Non availability of exotic varieties	63	82.89
7	New recommended practices are tedious	54	71.10
8	Recommended practices are not considered appropriate by the locals	60	78.94
9	Traditional bee rearing methods are better than the modern recommended methods	51	67.11
10	No extension activities organized for teaching the recommended hive management practices	0	00.00

There were some suggestions given by the beekeepers to increase honey production as well as to improve the adoption level of recommended beekeeping practices. Some of the suggestions are to integrate agroforestry practices in order to increase pollination among the bee. The farmers also stressed upon regular organization of training session to introduce the farmers with the latest beekeeping and beehive technologies.

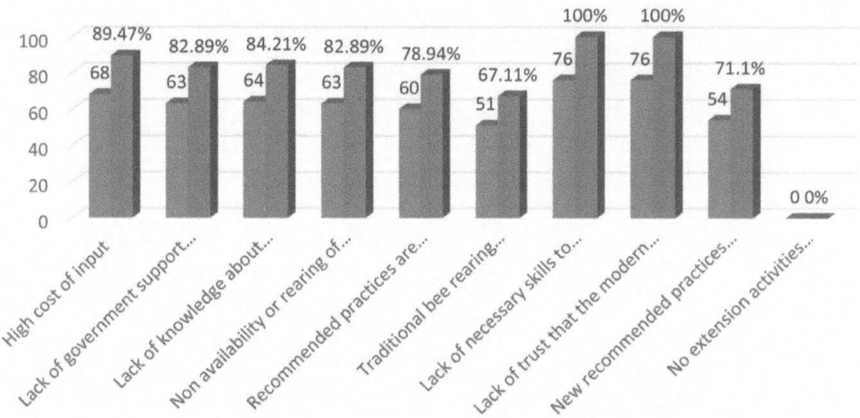

Fig. 1.0: Constrains faced by the beekeepers in adoption of recommended beekeeping practices

Conclusion

In the present study conducted the Beekeepers were observed to have a significant extent of about general aspects, management of boxes and bee flora but had a low-extent of adoption in some important areas like bee breeding, queen management, bee enemies and disease management which are important for booming the apiculture Industry.Therefore, the concerned agencies like State Beekeeping Research Centre, KVKs of the districts and other NGOs should guide the beekeepers and conduct more and more skill development training programs for beekeepers so as to overcome these lacunae. Certain serious steps are required to be taken by the government to provide support for the prices and reduce high cost of inputs involved in Apiculture. The beekeepers should be encouraged to believe that the modern recommended practices are useful and it can aid in positive Income generation.

References

Afreen Mallik. 2019. Effectiveness of Participatory Newsletter on Honey Production: A Study in Nainital District of Uttarakhand. Thesis, Doctor of Philosophy, G. B. Pant University of Agriculture and Technology, Pantnagar, Uttarakhand, India. pp 12-13.

Khan, M.S., Karnatak, A. K. and Srivastava, P. 2007. Beekeeping status in Uttarakhand. Indian Bee Journal., 69(1/4): 88-93.

Tiwari, P. and Singh, D. 2010. Status of Beekeeping in District Chamoli and Rudraprayag of Garhwal Himalaya, Uttarakhand. HMNEH souvenir organized by the state beekeeping center Jeolikote.

Yogesh Kumar. 2013. A multi-dimensional study on production and management system ofapiculture farming in Jammu region. Thesis, Doctor of philosophy, Shere-e-Kashmir University of Agriculture Sciences and Technology of Jammu, India.

24

Agro-Entrepreneurship A Way to Boost Startup Ecosystem in Agriculture

Lade A.H.[1], Ahire R.D. [2] and Shinde, G.U.[3]

[1]Department of Extension Education, COA, VNMKV, Parbhani, Maharashtra
[2]College of Agriculture, Badnapur, Mahrashtra
[3]NAHEP-DFSRDA-ICAR, VNMKV, Parbhani, Maharashtra

Introduction

Agriculture is one of the important pillars of the Indian economy. According to a report from FICCI, about 54 percent of Indian population depends directly on agriculture and it accounts for around 17.3 percent of GDP (FICCI, 2018). In the last two decades, powered by the Government policies and strong engagement of the industry and institutions, agriculture is rapidly evolving into agribusiness in terms of approach and structure. Transformation of Agriculture to Agri-business is one of the important strategies where enterprising farmers practice profitable agriculture.

India has become the third largest startup ecosystem hub. India is home to highest number of unicorn startups after US and China with 26 unicorns out of 250 plus total unicorns globally. At a time where with the increasing population and demand for better quality and higher quantity of food. India currently has more than 49,000 plus startups, 1500 plus investors, 250 plus incubators, and 26 unicorns (NASSCOM, Srivastav, 2018). The Indian startup ecosystem is ranked at number 37 and the global ranking of its top three cities are Bengaluru, New Delhi and Mumbai for the year 2018. In 2020, Bengaluru again continued its title "silicon valley of India" (Spews, 2021).

In a bid to double the farmer's income by 2022, the Government of India is continuously looking for ways to boost agricultural production, food processing and marketing avenues through the integration of latest technologies and innovations; thus creating a huge scope for food and agritech startups in the country (Balaji, 2018). Many agritech startups in India are mainly in marketplace segment where e-commerce companies provide fresh and organic fruits and vegetables procured directly from farmers. Very recently many startups have come up providing innovative and sustainable solutions for farmer's problems. Startups have provided solutions such as biogas plants, solar powered cold storage, fencing and water pumping, weather prediction, spraying machines, seed drills, vertical farming, etc (Sachitanand, 2018). EM3 Agriservices offers farming services and machinery rentals to farmers on a pay-for-use basis.

Agri start-ups have received decent support from the government through supportive policies such as Start up India, Atal Innovation Mission, New Gen Innovation and Entrepreneurship Development Centre, the Venture Capital Finance Assistance (VCA) Scheme promoted by the Small Farmers' Agri-Business Consortium and ASPIRE scheme.

Agritechis the idea of applying modern technologies to the agricultural sector with a view to enhance produce, efficiency and revenue. The Indian food and grocery market is the world's sixth largest, with retail contributing 70% of the sales. The Indian food processing industry accounts for 32% of the country's total food market, one of the largest industries in India and is ranked fifth in terms of production, consumption, export and expected growth.

What is Startup?

Startup is a newly established business (Oxford Dictionary).

Startupis "a team of entrepreneurial talent developing new innovations, in identifiable and investable form, in progress to validate and capture the value of the created innovation - with ambition to grow fast with scalable business model for maximum impact.

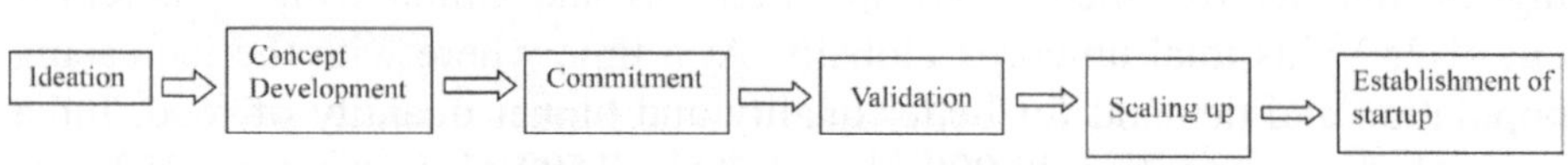

Key features of development stages of startup

Stages	Name	Key Features of Development
Stages I	Ideation	• Entrepreneurial ambition of the founder(s) • Potential development of the product or service idea for a big enough target market • Some initial business models and ideas on how the idea would create value or make money • Presence of single person or a vague team • No confirmed roles or commitment in the team structure yet.
Stages II	Concept development	• Defining of mission and vision of the startup with initial strategy • Setting up of key milestones and goals for next few years • Formation of core team consisting of co-founders with complementary skills and ownership plan and rights • Inclusion of some additional members for the requirement specific skill sets along with ownership
Stages III	Commitment	• Committed skills balanced founding team sharing same vision, values and attitudes • Building the MVP or Minimum Viable Product for the users to test their business idea • Signing of SHA: Shareholder Agreement among the co-founders It includes milestones that need to be achieved • Commitment of time and money shared by each shareholder • SHA signed for next three years with vesting terms
Stages IV	Validation	• Most important stage from the point of view of the founder(s), employees, initial customers and angel investors through initial user growth • Initial Key Performance Indicators (KPIs) identified • Founders are struggling to find the right product strategy & brand positioning that would allow them to attract potential Series A/B venture investment • Most of the startups lose their plot during this stage of business.
Stages V	Scaling up	• Focus on KPI based growth in terms of users, customers and revenues or market share in the target market • Potential to grow fast Received series A funding for the startup • Maximum time involved in hiring resources, improvement and distribution of product to target population and implementing new or established process

Stages	Name	Key Features of Development
Stages VI	Establishment of startup	• Have achieved great growth that can be expected to continue • Easy to attract funding and customers now Depending on vision, mission and goals will continue to function "like a startup" • Founder(s) or Investors may decide to exit or continue with the company

Source: Aggarwal, 2015; Segal, 2016; Yasuda, 2016

An Entity Shall Be Considered as A "STARTUP" When It Fulfills the Criteria

- Period of existence and operations should not be exceeding 10 years from the Date of Incorporation
- Incorporated as a Private Limited Company, a Registered Partnership Firm or a Limited Liability Partnership
- Should have an annual turnover not exceeding Rs. 100 crore for any of the financial years since its Incorporation
- Entity should not have been formed by splitting up or reconstructing an already existing business
- Should work towards development or improvement of a product, process or service and/or have scalable business model with high potential for creation of wealth & employment

(Source :Aadithya, 2020)

Top Ten Startup Hubs in India

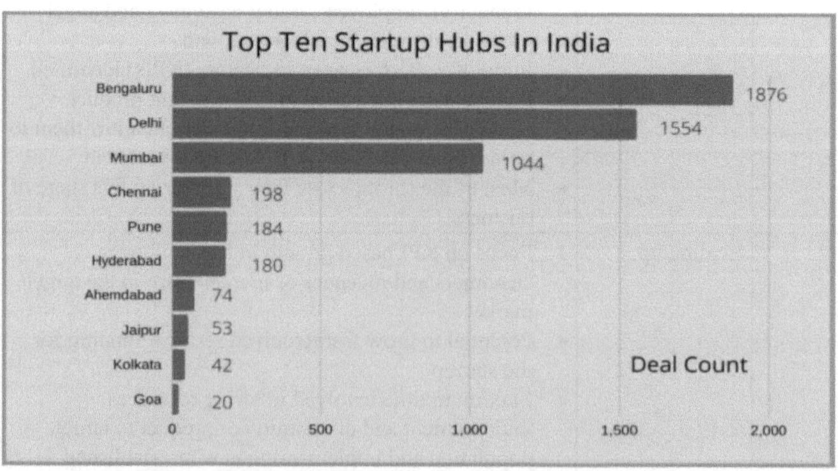

(Source: Startuptalky, 2020)

State-wise focus on agri start-ups in terms of number (2013–17)

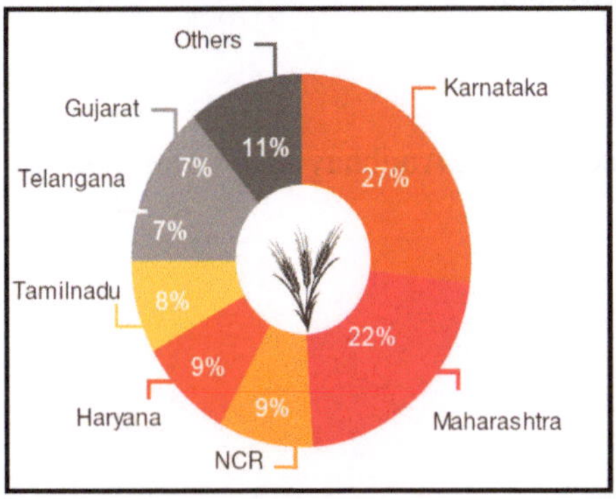

State-wise status in terms of funding (2013–17)

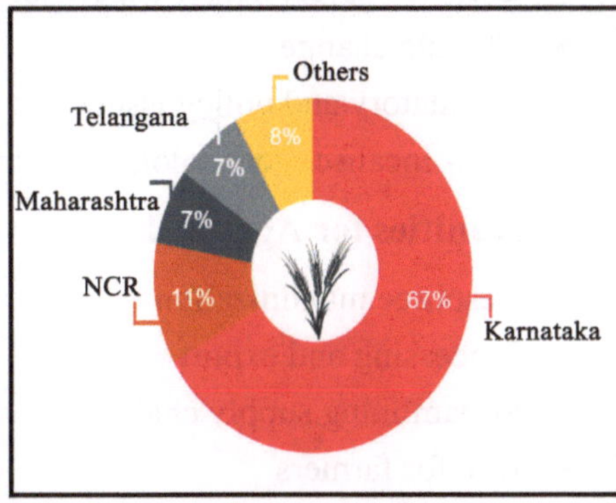

Source: Agritech in India – maxing India farm output, NASSCOM (June 2018)

Broad categorisation of agri start-ups based on solutions offered in the value chain	
Big data	Using farm data to determine opportunities and key areas • Farm management solution • Risk mitigation and forecasting solution • CRM and input channel solution • Traceability and compliance
Market linkage model	Helping farmers to keep abreast with market prices and scenario • Agri inputs market platforms • Real-time solution for farmers • Updated agriculture information • Quality, availability and price checks • Farm to fork supply chain
Farming as a service	Providing affordable technology solutions for efficient farming • On-demand harvesting • Digital payments • Market pricing • Enabling technology to reach farmers • Agricultural machinery platform
IoT-enabled technology	Using IoT devices for remote monitoring and tracking • Vertical farming monitoring solutions • Hydroponic farming ecosystem to monitor humidity, air temperature, etc. • Aeroponics system for smart farming

(Source: FICCI, 2018)

Challenges & Problems in Agri start-ups

- Insufficient supply chain
- Lack of financing
- Inadequate irrigation

- Farm size vs Productivity
- Lack of subject matter experts or mentors:
- Climate change
- Regulatory and policy issues
- Non- localised technologies(Source:Vikash, Anubhav and Kiran, 2021)

Opportunities for Agri-tech

- Resource maximization
- Processing and export
- Streamlining supply chain
- IOT for farmers
- Employment rationalization
- Fintech for Farmers (Source:Vikash, Anubhav. Kiran, 2021)

Schemes Initiated for Agri Start-ups

Sr. No	Name of scheme	Silent Feature
1	Start Up India	Startup India is a flagship initiative of the Government of India, which aims to build a strong ecosystem for nurturing innovation and start-ups in the country, to drive sustainable economic growth and generate large-scale employment opportunities
2	Atal Innovation Mission (AIM)	Atal Innovation Mission (AIM) including Self-Employment and Talent Utilization (SETU) is the Government of India's endeavour to promote a culture of innovation and entrepreneurship. AIM provides a grant-in-aid of 10 crore INR to each Atal Incubation Centre for a maximum of five years to cover the capital and operational expenditure cost in running the centre.
3	NewGen Innovation and Entrepreneurship Development Centre (NewGen IEDC	The government's NewGen IEDC startup Program is implemented in educational institutions. A maximum of 20 new projects are supported in a year and the government provides one-time, non-recurring financial assistance, up to a maximum of 25 lakh INR to the institution for the establishment cost, furnishing of cubicles for start-ups, purchase of PCs with printers, library books, journals, laptop, multimedia projector, 3D printers etc.
4	Venture Capital Finance Assistance (VCA)	Capital assistance will depend on the project cost, location and the promoter's status. It will be in the form of interest-free venture capital assistance up to 50 lakh INR or 26% of the promoter's equity, whichever is lower.

(*Source* ; Anupam Anand. Saravanan, 2019)

Agri startup in India

- BigHaat is founded by a team of avid entrepreneurs in the year 2015. BigHaat is India's largest Agri Inputs Marketplace Platform providing wide choice of quality inputs to farmers at their doorstep.
- **AgroStar was founded in April 2013** by Shardul Sheth and Sitanshu Sheth. It is an mcommerce platform for agriculture and has built a technology platform and uses data analytics and agronomy expertise to provide real-time solutions to farmers in India. It is a direct-to-farmer digital platform where farmers can procure quality agri inputs at fair price just by using a mobile phone.
- BharatAgri was founded in 2017 by Sai Gole and Siddharth Dialani. BharatAgri offers farmers an app-based platform for AI-based agronomy services on a paid subscription basis, which increases farm incomes through systematic implementation of scientific farming technique. BharatAgri's proprietary algorithm collects data on more than 30 critical parameters to provide customized advisory across multiple crops, including grains, pulses, fruits, and vegetables. Farmers get a fully digital and automated service that provides targeted insights at appropriate times and regular monitoring of crops.

Conclusion

- Startups are working tremendously in changing the face of Indian agriculture by providing farmers with low-cost farming solutions
- Majority of the startups fail after sometime due to lack of ideas to better their product, process or services. So it's very simple math that if a startup wants to stay in competitionit needs to adopt a culture of innovation in its DNA
- The Government of India is continuously looking for ways to boost agricultural production, food processing and marketing avenues through the integration of latest technologies and innovations; thus creating a huge scope for food and agritech startups in the country

References

Aadithya P. S. (2021) Blooming Era of Indian Agritech Start-ups, just agriculture multidisciplinary e-newsletter, (2):147-152.

Aggarwal P (2015) 6 stages of Digital Startup business. Retrieved from http://piyushaggarwal.me/ stages-of-digital-startup-business/#.XFCNYFUzbZ.

Agri-start-ups-Knowledge-report- 2018-ficci p0p.

Anupam Anand. and Saravanan, R. (2019) Agritech Startups: The Ray of Hope in Indian Agriculture, Discussion Paper 10, MANAGE-Centre for Agricultural Extension Innovations, Reforms and Agripreneurship, National Institute for Agricultural Extension Management (MANAGE), Hyderabad, India.

Balaji S (2018) India's farmers could be making more money soon, with the help of clever agritech startups. Retrieved from https://www.forbes.com/sites/sindhujabalaji/2018/03/19/indias-farmers-could-be-making-more-money-soon-with-the-help-of-clever-agritech-startups/#5cedc4bd7abe.

https://research.agfunder.com/2016/AgFunder-Agtech-Investing-Report-2016.pdf.

NASSCOM (2018) AgritechIn India – Maxing India Farm Output. Retrieved from https://www.nasscom.in/knowledge-center/publications/agritech-india-%E2%80%93-maxing-india-farm-output.

Srivastav S (2018) Indian Startup Ecosystem: A Global Growth Story in The Making. Retrieved from https://inc42.com/features/indian-startup-ecosystem-a-global-growth-story-in-the-making.

Startuptalky. (2020). The Growth of Startup Ecosystem In India. Startuptalky, Retrieved from The Growth Of Startup Ecosystem In India (startuptalky.com).

Vikash, Anubhav and Kiran (2021) Agri Start-up in India: Opportunities and Performance in Indian Context, just agriculture multidisciplinary e-newsletter, (1):1-13

25

Developing Entrepreneurial Qualities Amongst the Member Farmers Through FPOs

Swati S. Khandave[1], V.J.Tarde[2], Jyoti Deshmukh[3]

[1&2]Department of Agricultural Extension Education, College of Agriculture Pune, Maharashtra
[3]Department of Agricultural Extension Education, College of Agriculture Latur, Maharashtr

Abstract

Indian agricultural scenario reflects that 85.00 per cent of the farmers are small and marginal having less than 2 hectares of land holding. As this group is operating at subsistence level, it is necessary to develop entrepreneurial qualities among them to improve their income.In order to meet the dream of Atmanirbhar Bharat and to make farmers atmanirbhar, Farmer Producers' Organizations (FPOs) is an appropriate form to mobilize these farmers through decentralized production and centralized services. Recently Government of India has announced a budgetary provision of 4496.00 crore for five years for promotion of 10,000 FPOs.(Cabinet Committee on Economic Affairs,CCEA2020)

An attempt was made to study the entrepreneurial attributes of the FPO members,to know personal,socio economical characterstics its influence on entrepreneurial behavior and constraints faced by them in enterprising. On the basis of normalized rank order method as recommended by Guilford (1954), the rank values for each attribute was worked out.

Based on maximum number of FPOs, major three tahsils viz., Khed, Junnar and Purandar of Pune district from Western Maharashtra and five, four and two FPOs engaged in vegetable production were selected, respectively by using multistage sampling procedure. Further based on number of members and villages covered, 10 per cent of sample was drawn from each FPO by proportionate sampling method, constituting total sample of 200 respondents.

The results of the study revealed that great majority(95.41%) of the members expressed that market orientation is the most relevant and important attribute with its rank value7.44 and assigned 1st rank followed by decision making ability (6.99) and economic motivation (6.42) stated by vast majority(94.29 % and

92.85%) judges with 2nd and 3rd rank respectively. Also study further stated that planning ability (6.09),achievement motivation (5.81), self confidence (4.66), scientific orientation (4.45) were the important qualities with rank 4,5,6,7 in descending order respectively. Study further stated that lack of share capital of an organization, lack of guidance and support from resource institution, lack of co-ordination and managerial skills among the board of directors(BoD), lack of basic infrastructure facilities etc. were the major constraints for developing an enterprise.

As present study rightly explored the attributes of entrepreneurship of FPO members it can be helpful to increase investments in business and administrative capacities of members so as to make organizations more competitive functional units. It is important to resolve the constraints faced members to make organisations sustainable.

Keywords: *FPO, BoD, Entrepreneurs, Resource institutions*

Introduction

In view of the present Indian agriculture scenario, farmer producers organisations (FPOs) are emerged as a most appropriate institutional form to mobilize small and marginal farmers ,improved their access to technology, inputs and markets. As per the report of National Council on Farmers, these organisations are mainly promoted to combine the advantages of decentralized production and centralized services, post harvest management, value addition and marketing. (NCF,2004). In order to meet the dream of Atmanirbhar Bharat and to make farmers atmanirbhar, Farmer Producers' Organizations (FPOs) is an appropriate form to mobilize them. Recently Government of India has announced a budgetary provision of 4496.00 crore for five years for promotion of 10,000 FPOs.(Cabinet Committee on Economic Affairs,CCEA2020)

Also to improve agricultural productivity and income of the farmers, it is necessary to develop an entrepreneurial culture among the member farmers of an organisation. To make this organisational approach more successful, member farmers need to develop planning ability, decision making abilitiy, managerial qualities ,scientific attitude, organisational skills, participatory skills etc. The member farmers as an agripreneur should possess knowledge about improved production techniques, markets, consumer preferences , schemes, credit availability, policy advocacy to enterprise effectively under FPO.

Developing entrepreneurship among the members is the complex process which includes many factors or indicators that are interrelated and interdependent. Entrepreneurs are not required to cause the change but exploit the opportunities that change creates in technology, market demand, social norms, etc. In present research study, an attempt was made to study the entrepreneurial attributes of the FPO members,to know personal,socio economical characterstics its influence on entrepreneurial behavior and the constraints faced by them in enterprising.

Methodology

The possible attributes/indicators of entrepreneurship development were identified after reviewing the related literature. After thorough scrutiny and comprehensive understanding, these indicators were consulted with the group of experts,thinkers, educationists from SAUs.

Those indicators found relevant by more than 75 per cent of the judges were considered as relevant items. Based on relevancy and rank assigned finally ten major entrepreneurial attributes were considered, the sub-items of each indicator were also finalized. In order to study the entrepreneurial attributes of members and their personal,socioeconomical,situatonalcharacterstics and its influence on entrepreneurial behaviour Pune district was purposively selected. Based on maximum number of FPOs, major three tahsils viz., Khed, Junnar and Purandar of Pune district from Western Maharashtra and five, four and two FPOs engaged in vegetable production were selected, respectively by using multistage sampling procedure. Further based on number of members and villages covered, 10 per cent of sample was drawn from each FPO by proportionate sampling method, constituting total sample of 200 respondents.

The members were requested to give their valuable opinion regarding the relevancy in developing entrepreneurship and assign rank as per their importance. The entrepreneurial behaviour index was calculated and respondents were grouped in to three categories to study influence of entrepreneurial behavior on personal profile of member. The results obtained were presented as below.

Results and Discussion

Table 1: Entrepreneurial attributes of FPO members and its relevancy (n = 70)

Sr. No.	Item	Relevance			
		Frequency	**Percent**	**Rank value**	**Rank**
1	**Market orientation**				
	• Market availability				
	• Market demand				
	• Consumer needs and preferences				
	• Direct marketing under brand name	67	95.71	7.44	I

Sr. No.	Item	Relevance			
2	**Decision making ability**				
	• Timely decisions				
	• Prompt implementation				
	• scientific and rational decision to join FPO, to purchase inputs, participate in FPO extension activities	66	94.29	6.99	II
3	**Economic motivation**				
	• Remunerative price to produce				
	• Reduce cost of production				
	• More economic returns				
	• Improve bargaining power and economical status	65	92.85	6.42	III
4	**Planning ability**				
	• Advanced planning for purchase of inputs ,labour ,credit				
	• Timely planning to manage resources, farm operations, marketing of produce	64	91.42	6.09	IV
5	**Achievement motivation**				
	• Hardship to achieve goals,				
	• Endless efforts to reach objectives and gain success				
	• Accept challenges	63	90.00	5.81	V
6	**Self confidence**				
	• Positive attitude, Self identity				
	• Confidence about use of improved technology, marketing ,success in FPO business activities	62	88.57	4.66	VI
7	**Scientific orientation**				
	• Encourage systematic cleaning, grading and scientific storage				
	• Scientifically grown vegetables fetch better price				
	• Rapport with KVK/SAU scientists	60	85.71	4.45	VII
8	**Innovativeness**				
	• Use of new varieties/improved technique,				
	• New business opportunities .marketing strategies				
	• Service competitiveness	56	80.00	3.29	VIII

Sr. No.	Item	Relevance			
9	Communication skill				
	• To develop organizational relations				
	• Enable team building				
	• Help to solve group disparities/ problems Need for linking business partners	55	78.57	1.59	IX
10	Risk orientation				
	• Work in uncertainty condition				
	• Trying new production technique/ method	54	77.14	1.09	X

Data from Table 1 revealed that great majority(95.41%) of the judges expressed market orientation is the most relevant and important indicator with its rank value7.44 and assigned 1st rank followed by decision making ability (6.99) and economic motivation (6.42) stated by vast majority(94.29 % and 91.43%) judges with 2nd and 3rd rank respectively. Also judges stated that planning ability (6.09), achievement motivation (5.81), self confidence (4.66), scientific orientation (4.45) were the important qualities needs to desire by the FPO member with rank 4,5,6,7 in descending order respectively. Further majority of the expert rated innovativeness (2.28), communication skills (1.59) risk orientation (1.09) with rank 8th,9th and 10th in descending order respectively. Similar results were reported by Chaurasiya *et al.* (2016),Gamit et al .(2015),Singh A.P.(2013).

It is inferred from above data that market orientation, economic motivation, decision making ability were the major determinants needs to possess by FPO members to enterprise successfully, followed by planning ability, achievement motivation, scientific orientation and self confidence next important determinants for proper management of recourses, developing rapport with scientists for use of improved technology as rated by judges.

Table 2: Personal, social, economic and situational characteristics of member farmers

Sr.No.	Particulars	Frequency	Percent	Mean of E.B.I.
Personal Characterstics				
A.Age				
1	Young (18 to 35 years)	81	40.50	80.32
2	Middle (36 to 50 years)	88	44.00	78.55
3	Old (Above 50 years)	31	15.50	75.28
B.Education				
1	Illiterate	2	1.00	74.93
2	Functionally literate	6	3.00	75.50
3	Primary education	20	10.00	76.84

Sr.No.	Particulars	Frequency	Percent	Mean of E.B.I.
4	Secondary education	39	14.50	77.81
5	Higher Secondary education			
	(11th to 12thor any Diploma course)	49	24.50	78.78
6	Graduation/Post-graduation	84	42.00	80.62
Social characteristics				
C.Occupation				
1	Agriculture Only	96	48.00	77.48
2	Agriculture + Horticulture	36	18.00	78.70
3	Agriculture +Horticulture +Animal husbandry	28	14.00	81.53
4	Agriculture + Service	14	7.00	79.13
5	Agriculture + Business	26	13.00	79.38
	D.Social participation			
1	Membership in one organization	66	33.00	77.85
2	Membership in more than one organization	87	43.50	78.89
3	Office bearer	47	23.50	80.05
Economic characteristics **E. Land holding**				
1	Marginal (Up to 1 ha.)	80	40.00	79.21
2	Small (1.1 to 2 ha.)	87	43.50	78.26
3	Semi-medium (2.1 to 4 ha.)	25	12.50	78.51
4	Medium (4.1 to 10 ha.)	08	04.00	79.57
F.Annual income				
1	Low (Up to .1,18,200)	54	27.00	78.01
2	Medium			
	(.1,18,201 to .3,87,000)	104	52.00	78.50
3	High(Above .3,87,000)	42	21.00	80.11
	G. Credit utilization			
1	Procurement of agriculturalinputs	Low (upto .0.5 lakh)	113	56.50
2	Purchase of machinery, electric motor ,preparing farm pond, etc.	Medium		
		(.0.5 to .1.0 lakh)	62	31.00
3	Purchase of tractor, rotavator, land leveling, irrigation set etc. (more than 1.0lakh)	High	25	12.50
Situational characteristics				
H. Irrigation methods				
1	Flood	35	17.50	78.89
2	Ridge and furrows	33	16.50	78.45

Sr.No.	Particulars	Frequency	Percent	Mean of E.B.I.
3	Flood+Sprinkler	28	14.00	78.49
4	Flood+ Drip	47	23.50	78.72
I. Cropping pattern				
1	Fruit vegetables – wheat – onion	126	63.00	II
2	Fruit vegetables – potato – green gram	114	57.00	III
3	Leafy and fruit vegetables – cole crops – groundnut	142	71.00	I
4	Onion - chick pea - fodder crops	85	42.50	IV
5	Okra - green peas – gourds	60	30.00	V
J.Market distance				
1	Short (below 27km)	45	22.50	79.43
2	Medium			
	(27 to 67 km)	118	59.00	78.46
3	Long (above 67 Km)	37	18.50	78.90

The data from table 2 indicated that more than two fifth (44.00 per cent) of the respondents were found in middle age group whereas 40.50 per cent were in young age group followed by 42.00 per cent of respondent had completed graduation and post graduationdegree ,nearly one forth (24.50 per cent) were completed higher education .

As regards to occupation,nearly half (48.00 per cent) of the members engaged in agriculture whereas, 18.00 per cent respondents engaged in cultivating agriculture + horticulture crops. Data regarding social participation indicates that slightly more than two-fifth (43.50 per cent) respondents were having medium membership i.e. member in more than one organisation whereas one third (33.00 per cent) of the respondents were members of one organization indicating low membership.

Economic characteristics viz.annualincome,land holding and credit utilization were studied. The data showed that majority (83.50 per cent) of the respondents were small and marginal farmers with annual income ranging from .1,18,201 to .3,87,000(52.00 %) followed by 27.00 per cent had low annual income i.e.,upto .1,18,200 lakh. Further it is observed that

more than half (56.50 per cent) of the members availed credit less than .50,000 and were utilized as a crop loan for purchasing inputs viz.,seeds, fertilizers, plant protection chemicals, vegetable carets,etc. whereas 31.00 per cent availed credit upto .100,000 and utilized for purchase of equipments viz., sprayers, dusters, preparing farm pond etc.

Data pertaining to situational variables revealed that comparatively more (28.50 per cent) number of farmers adopted ridge and furrows with drip method of irrigation, whereas 23.50 per cent had adopted flood as well as drip method combine based on water availability and followed Leafy- fruit vegetables-cole crops-groundnut cropping pattern (71.00%) with first rank. The cropping pattern fruit vegetables-wheat-onion was adopted by 63.00 percent respondents with second rank. Majority (81.50 per cent) of the respondents reported that market is located at medium to short distance from collection centre.

It is clear from the table that entrepreneurial behaviour index was substantially increased with increase in age upto certain period, education, occupation, socialparticipation, annual income means these characteristics had shown positive influence on developing entrepreneurial behaviuor of members. Similar findings were observed by Thakur (2015),Savitha (2014). However in case of land holding,credit utilization, irrigation methods,market distance the entrepreneurial behaviour index of all categories was almost similar indicating no influence on entrepreneurial behavior. These variables may change as per varying situations hence might have not shown influence on entrepreneurial behavior.

Entrepreneurial behaviour of the member farmers

In order to study the entrepreneurial behaviour of the members of an organisation, maximum score obtained by the member for all entrepreneurial attributes was considered and their entrepreneurial behavior index was calculated. Based on SD and mean members were categorized into three categories as below.

Table 3: Distribution of the respondents by their entrepreneurial behavior index

(n = 200)

Sr.No.	Category	Frequency	Percent
1	Low (upto73.96 E.B.I.)	27	13.50
2	Medium (73.97to 82.56 E.B.I.)	135	67.50
3	High (above 82.56 E.B.I.)	38	19.00
	Total	200	100.00
X = 78.76 S.D. = 4.80			

It can be seen from Table 3 that more than three fifth of the respondents (67.50 per cent) were observed in the medium level of entrepreneurial behaviour index, while nearly one fifth(19.00 per cent) respondents fell under the category of high entrepreneurial behaviour. The remaining 13.50 per cent respondents possessed low level of entrepreneurial behaviour. Similar findings were reported by Chaudhary (2017).

The important constraints reported by the members were lack of share capital of an organization, lack of guidance and support from resource institution, lack of co-ordination and managerial skills among the board of directors, lack of basic infrastructure facilities etc. which needs to address properly.

Conclusion

As a result of continuous support from state as well as central government, number of FPOs has been substantially increased in last decade throughout the country. But to make these organizations more competitive functional units it is important to develop certain skills and qualitative attributes among the members. Present study rightly explored the qualitative attributes of entrepreneurship which will be helpful to the extension agencies, incubation centres of FPO in planning and selecting target group for training, promoting FPOs for participatory and collective activities. These findings also useful to the board of directors members of FPO for developing administrative and managerial capacities for increasing investments in business so as to make the unit more competitive and economically stable.

References

Anonymous, (2004). Report of National Commission of Farmers, New Delhi.

Cabinet Committee on Economic Affairs, (CCEA) (2020). CCEA Approves scheme for Formation and promotion of to form and promote 10,000 new FPOs,Pressrelease,Press Information Bureao. Government of India.

Chaudhary, K.L.(2017). Entrepreneurial Behaviour of Dairy farmers in North Gujarat. Ph.D.Thesis (Unpublished). Sardarkrushinagar Dantiwada Agricultural University, Sardarkrushinagar.

Chaurasiya, K.K.; Badodiya, S.K.; Somvanshi, S.P.S. and Gaur, C.L. (2016.) Entrepreneurial behavior of dairy farmers in Gwalior district of Madhya Pradesh, Indian Journal of Dairy Science, 66(1):112-115.

Gamit, M. P., Durgga Rani V., I. N. Bhabhor, K. K. Tyagi, A. D. Rathod. (2015) Entrepreneurial behaviour of dairy farmers in Surat district of South Gujarat International Journal of Advanced Multidisciplinary Research 2(8):50–56

Guilford, J.P. (1954). Psychometric Methods, Tata McGraw Hill Book Publication Co. New Delhi. : 178-183.

Savitha, C.M., Siddaramaiah, B.S. and Shankara, M.H. (2014).Factors contributing to entrepreneurial behaviour of rural and urban women., International Journal of Farm Sciences 5 (2): 217-223.

Singh, A. P. (2013). Factors influencing entrepreneurship among farming community in Uttar Pradesh.International refereed Research Journal .4(3): 115-117.

Thakur, A. Barman, U. (2015). Determining Predictors Influencing Entrepreneurial Behaviour of Women Self Help Group Members. Indian Research Journal Extension Education.15(3):52-55

26

Good Practices in Agricultural Wholesale Markets: Answer to COVID 19 Challenges

Anil Chauhan[1], Kiran[2], Amit Kanawjia[3]

[1]*SMO,State Agricultural Marketing Borad, Khalini, Shimla Himachal Pradesh*
[2]*Genetics & Plant Breeding, CSKHPKV, Palampur, Himachal Pradesh*
[3]*Horticulture, KVK, Jalaun, BUAT, Banda, Uttar Pradesh*

Abstract

Agricultural wholesale markets are import component of agricultural output supply chain. In India most of agricultural wholesale trade is done in yards of Agricultural Produce Market Committees (APMCs). These markets are regulated through Agricultural Produce Marketing Acts of states. During the COVID 19 period, these wholesale markets are playing a very crucial role in maintaining the supply chain. In most of states the markets remain operational during the period of crises. Wholesale markets adopted many innovative steps to prevent the spread of virus. The COVID 19 has affected the whole agricultural supply chain system.It has affected the present income of farmers and some of its impacts will be visible in future. In this article various challenges faced by wholesale markets, good practices adopted in these markets, steps taken by the government at Central State level, future challenges and suggestions are mentioned.

Keywords: *Good Practice, COVID 19, Food Grain, Market Yard, Challanges Market*

Introduction

Novel Corona Virus 2019 (COVID 19) has impacted the human race very badly than any other disease in twenty first century. Almost every country on this planet was affected by this virus. The condition is more severe in developed countries.

So when it All Started?

There were reports that the medical professionals of Wuhan in China informed the government that there is some mysterious pneumonia like disease which is spreading very fast and affected the doctors and other paramedical staff who are attending the patients. When the situation became serious it was declared as outbreak of epidemic and first case of COVID 19 was declared in Wuhan China on 31st December 2019. Wuhan is one of the main business centres of china and there was regular movement of domestic and international visitors to that place. Till the problem was identified and communicated to other agencies, the spread of virus was already there in other parts of china and world specially Europe and USA.

World Health Organisation (WHO) declared COVID 19 as pandemic on 11 February 2020. Countries like USA criticized WHO for delay in identifying the disease and declaring it a pandemic.The damage was already done due to its spread in other countries. In India first reported case of COVID 19 was on 30th January at Kerala. However, considering the possible danger of this disease, Indian Government took primitive measure to contain the spread of Virus and announced Nationwide Lockdown 1.0 on 23rdMarch 2020. Similarly, Lockdown 2.0, 3.0, 4.0 were announced on 14 April, 3rd May and 17th may respectively.

During the lockdown process all the commercial establishments were closed and only essential services were operational. It was the duty of District Administration to fulfilthe basic requirements of public like foodgrains, milk, vegetables, medicines etc. It was proposed that there should be minimum human touch with the consumable items. After some days the agricultural services were permitted to operate and permission to open the wholesale markets were granted. At this time the rabi crops were almost ready to harvest. Wheat which is a major rabi crop is ready to harvest and farmers were planning to arrange the resources for its post harvest management. In Some hot areas harvesting was already done. Machinery and labour were the major resources which need to be mobilized during that period. Due to relaxation in lockdown norms there is partial movement of these two resources, which helped the harvesting process.

In the 3rd advanced estimates of production by Ministry of Agriculture and farmers welfare, there will be record production of crops in this year. As per 3rd Advance Estimates, the estimated production of major crops during 2019-20 is as under:

Foodgrains: 295.67 million tonnes. (record)

Rice – 117.94 million tonnes. (record)

Wheat – 107.18million tonnes. (record)

Nutri / Coarse Cereals: 47.54 million tonnes. (record)

Maize – 28.98 milliontonnes. (record)

Pulses – 23.01 million tonnes.

Tur – 3.75 million tonnes.

Gram – 10.90 million tonnes.

Oilseeds: 33.50 million tonnes. (record)

Soyabean – 12.24 million tonnes

Rapeseed and Mustard –8.70 million tonnes

Groundnut – 9.35 million tonnes

Cotton: 36.05 million bales (170 kg per bale) (record)

Jute &Mesta: 9.92 million bales (180 kg per bale)

Sugarcane: 358.14 million tonnes

Challenges in Marketing

Availability of harvesting machinery and manpower

India is a diverse country. Numbers of crops are grown at different times in different parts of country, which require movement of harvesting machinery and manpower to perform the harvesting operations. The migration of agricultural labour is very common. At the time of lockdown, the labour and machinery was at different location from production areas. Their movement was a great difficulty during this time.

Hassel free movement of vehicles across the district and state boundaries

Some specific states or district have become epicentre of COVID 19 infections like Delhi and Mumbai. The movement of labour and vehicles to North and South India are through these major cities as these are the transport junctions. Due to outbreak of COVID 19 the state and district authorities has stopped the movement of transport which was a major challenge for marketing of agriculture commodities.

Preparedness of Market yards and collection centres by avoiding crowd

The main marketing centres of agricultural crops are markets operated by APMC and State Marketing Boards. Some of market yards and seasonal in nature means they operates only in harvesting season and remain closed for rest of year.Others are full time operational like fruits and vegetable markets of consuming areas. Majority of seasonal harvest come in seasonal yard for which preparation like repair and maintenance, arrangement of drinking water and sanitation, cleaning arrangement, repair and installation of grading machinery etc. are performed before the harvesting season. But due to lockdown at starting of harvest season has affected the preparation of these arrangements.

Maintaining social distancing and hygiene in markets

Agricultural markets are generally a crowded place as large number of stake holders visits these markets like farmers, labourers, transporters, weighmen, traders, commission agents etc. To operate these markets during COVID 19 period was very challenging by following the necessary norms like maintaining social distancing and proper hygiene during the operations of market.

Encourage minimum touch/ contact with agri./horti. Produce

The nature of spread of corona virus is through droplets of infected person. The virus in these droplets will remain live for long time on infected surfaces. Since all the post-harvest marketing operations were manual hence there were greater chances of infection at any stage of marketing. So to manage the minimum contact of harvested crops is great challenge during the operation of market.

Losses incurred in trade due to lockdown

Marketing of agricultural commodity is a business activity. Like any other business activity there is need of liquidity in market and significant capital is required to run the business. But due to lockdown the flow of liquidity was adversely affected and there are significant losses to tradersparticularly fruits and vegetables due to spoilage of crops in transit. Many of them have lost the capital which will run the future business. Hence the shortage of liquidity was very big challenge.

Slowing down of intercultural operations particularly horticultural crops like mango, grapes and apple

Due to non-availability of labour and agro chemicals the intercultural operation specially plant protection sprays were affected or recommended practices could not be adopted.Due to this there are chances that the productivity would suffer. This will have impact on demand, supply and prices in coming months.

Decrease in demand due to closure of hotel and other bulk buyers

The food servicing sector was closed during the lockdown. All the hotels, restaurants and other food points were closed. Due to this the demand for these bulk buyers wasalmost became zero, which affected the prices and value chain. The most severe category which was affected are farmers of high value crops and commercial crops. Marketing of these commodities became a very big challenge.

Availability of packing material

All the packing material manufactures were closed during the lockdown. The supply of raw material was also stopped. This affected the production and movement of ready material. The shortage of packing material was observed in areas where harvesting of crop was going on. This has also impacted the supply chain of packing material in area where crops in ready in next 2-3 months and large quantity of packing material is required.

Restriction on movement of traders

Traders play a very important role in price discovery of agricultural commodities. It is believed that the prices are high when there are large numbers of buyers. These buyers come from different parts of country to buy the produce. Due to spread of COVID 19 their movement was restricted which has affected the prices of commodities. This will also affect the future prices of commodities in markets as there are strict quarantine norms to move from one place to another.

Availability of supporting activities like auto repairs, spare parts, eateries hotel, lodges etc.

Although the movement of vehicles carrying agricultural produce was allowed during lockdown and curfew period but the supporting activities like repair and spare, hotels and dhabasetc were closed. This discourages the drivers and other functionaries to do the transportation activities.

Steps Taken to Address the Challenges of COVID 19

Central and state government has taken many stepsto manage the situation. Various micro innovations are happening at ground level depending upon necessity. Some of innovative steps are as under;

Steps taken at State Government and wholesale market level

- Agricultural Marketing department and marketing boards has advised Standard operating procedures (SOPs) for operations of whole sale markets. Emphasis was given to maintain proper social distancing and hygiene in wholesale markets. Recommended chemical was sprayed in premises of market on daily basis.
- Special marking of places in yard was made for vehicles and farmers coming to wholesale markets. Separate places were identified for farmer's vehicles and loader's vehicles. Separate timings were fixed for operation of both the activities.
- All the visitors of wholesale market were thermally scanned for any symptoms of COVID 19. If there were any flu like symptom, they were asked to leave the market and consult health officials. All the visitors were hand sanitized. All the vehicles coming to wholesale markets were also sanitized with recommended chemicals.
- It was made compulsory to wear mask in market. Face masks were provided to market functionaries like drivers, hamals, labourers' etc. To make aware about the various do's and don'ts to avoid spread of virus, marketing was done though public hoardings, flex hoardings, notice board messages and man to man contact.
- Some agricultural wholesale markets have installed hand free wash basins to enable the visitors to wash their hands frequently without touching the tap of bottle of sanitizer. These wash basins are foot operated and developed by at local level by fabricators.
- During the harvest time, permits were issued to labourers for crop cutting. They were directed to maintain social distancing and wear face masks.
- In some markets where there is huge arrival during marketing season,Token system was adopted to avoid rush. Farmers were told to dry the crop as per specifications and if moisture level is high, their produce will be rejected. This lead to extend the marketing season of crop and long queues were avoided. Some APMCs has introduced token system and desired numbers of farmers were allowed to enter the market. This avoided the crowding in market and helped in maintaining social distancing.

Steps taken at Union Government level

- Ministry of Agriculture and farmers welfare has launched Kisan Rath App on 17/04/20 to facilitate the logistics. This application is quite helpful to farmers for arranging the vehicle to carry the harvested produce to market. This is also helpful to the trader to carry the commodity from one market to other market.

- All India Transport Call centre (1800 180 4200 and 14488) was set up to facilitate logistic related problems. This number can be accessed from any phone from all over India. It helped the movement of vehicle seamlessly across the borders. This helpline coordinates between various state governments and solve any transport related issue.
- Kissan rails were operated by Indian Railways during lockdown period over 67 routes, which helped to maintain supply during the crises period.
- 100 new markets were connected with e-NAM and total number of wholesale markets linked with e-NAM are now 1000.
- During the COVID 19 period Union Government fixed aggressive procurement targets so that maximum crops which come to market can be purchased by Food Corporation of India, NAFED or State procurement agency. The numbers of procurement centres were also increased to 2790 from last year 1485

Challenges Ahead

Agricultural wholesale markets act as a very important component of Agricultural supply chain. On one hand it monetizes the farmers and on other hand ensures supply to consumers. A lot of market functionaries operate in these markets so these markets are more prone to COVID 19 infections. However, in current situation these markets have managed to curtain the spread except a few exceptions and are operating with almost full capacity. However, there are great challenges in future due to COVID 19 and its impact on agriculture and national economy as whole. Some of major challenges which can crop in future are as under:

- This pandemic has affected the resource movement. There is liquidity crunch in system, Traders has incurred losses, labour has migrated to native places and many more. So to streamline the system all the resources need to be arranged optimally to their specific position/location, which is a major challenge in future.
- Optimum storage utilization plays a key role to stabilization of prices and availability of commodity. Due to COVID 19, the commodity stored in cold stores, CA or MA Stores, godowns etc. Could not be moved due to lockdown. The challenge ahead is to move out the stored produce, maintenance of machinery and get it ready for storage of next crops. Highly efficient system of marketing and logistic support is required for it else there will be huge supply of commodity i.e. stored produce and next harvest which will affect the price and profitability.
- COVID 19 will affect the future decision making of farmers toward crops diversification like English vegetables, flowers etc. if new alternatives

or markets were not discovered for high value crops. These crops are consumed in hospitality industry and due to their closure, there is no buyer. Efforts of Government to promote diversification toward horticulture and other perishable crops will receive a setback as more and more farmers will plan to grow traditional crops like paddy and wheat to avoid risk and to get assured price through Minimum support price.

- International business is very badly affected due to COVID 19. Very high fluctuation in exchange rate has also changed the dynamics of export and import. Its impact will be commodity specific as some will get benefit and other will lose. However there will be significant shrinkage of international business and income level of household will decline which will affect the consumption. Unemployment level with rise due to possible recession which will affect the buying behaviour.

Suggestions

To meet the challenges posed by COVID 19 and its impact on Indian agriculture, some immediate and long term steps are required by Central and State Government. Some of suggestions are;

- A need was felt to constitute a national level task force comprising the representatives of Ministry of agriculture andfarmer's welfare, state marketing department / marketing boards and other stake holder who will plan the new strategy to meet the future challenges. This body can co-ordinate or operate the various function like Kisan rath app, National logistic support helpline, surplus deficit conditions etc.
- Decentralize the supply chain system and develop alternate centres which can operate in case of emergency situations. Dependence of one or two big markets needs to be changed.
- Re-assess the availability and utility of storage capacity keeping in view the production of agri commodities and their consumption.
- Big and liberalInvestment is required in agri-market infrastructure and its professional management
- Effective implementation and extra allocation of funds to Market intervention schemes to maintain market stabilization.
- Model on Credit in trade should be developed on e- NAM as most of agri supply chain system runs on credit and it is not available on e-NAM.
- Efforts should be made to increase interstate and intra state trade in e-NAM. This can be done through easy issuance of licenses within state or outside the state. A National level facilitation centre is required which can guide the traders.

- Developing stake holder specific training modules of e-NAM activities and trade. E.g. Farmer module, trader module, transporter module etc.

Social Distancing during harvesting

Social Distancing during Auctioning

Although Central and State governments are taking all the possible steps to meet the challenges of COVID 19 which will show its results in future. But immediate relief is required by farmer, labours and traders to meet the immediate challenge of livelihood. Efforts should be made to increase the real income of farmers. Facilitation should be given to trader so that size of value chain system can be increased, which will ultimately benefit every stakeholder and economy as a whole.

References

Acharya, S.S. and Agarwal, N.S. (1999), Agricultural Marketing in India, Oxford and IBH Publishing House, New Delhi.

Press information Bureaue, Government of India (2020), Third advance estimates of production of foodgrains, oilseeds and other commercial crops for 2019-20 [Press Release]. Retrieved from https://pib.gov.in/allRel.aspx

Chauhan, Anil (2016), Marketing of apple in shimla [Blog Post]. Retrieved from http://himachalhorticulture.blogspot.com/2016/

27

Role of FPOs in Uplifting Small and Marginal Farmers of Uttarakhand

Kiran Pant[1], A.K. Sharma[2], Ashima Muyal[3]

[1,2] Krishi Vigyan Kendra, Dedradun, Uttarakhand
[3] Banaras Hindu University, Department of Extension Education Varanasi, Uttar Pradesh

Abstract

India is bestowed with a lot of potentials to produce a variety of organic products due to its diversified agro-climatic regions and this is very true in case of Dehradun district of Uttarakhand. It offers a variety of geographic locations suitable for cultivating various food crops such as: fruits, vegetables, food grains etc. Agriculture and its allied sectors are the largest source of livelihoods in India. However, India still has many growing concerns. As the Indian economy has diversified and grown, agriculture's contribution to GDP has steadily declined from 1951 to 2011. (Source: http://www.fao.org/india/fao-in-india/india-at-a-glance/). Realizing these issues Government of India under the Department of Agriculture and Cooperation identified Farmer Producer Organization (FPO) as the most appropriate institutional form to mobilize farmers and build their capacity to collectively leverage their production and marketing strength(Ministry of Agriculture, 2013).For analyzing the potential, importance and contribution of FPO in the agriculture sector, the present study aims to discuss the role and impact of FPO in uplifting small and marginal farmers and its impact on the basis of the project undertaken by Silgur Bizat SwayattSahkarita FPO in Dehradun district of Uttarakhand. The present study revealed that the goals of FPO were achieved in terms of Utilizing Local agricultural Produce , developing Backward to Forward Linkages, Introducing Mechanical Processing for processing of vegetables and fruits, and developing Market linkages .The impact of the FPO was measured under six parameters viz: Livelihood Related Impact, Agricultural Development Related Impact, Migration Related Impact, Quality of Life Related Impact, Environmental Impact, Impact on Women Development with specific reference to gender issues, Convergence related impact and Impact of Capacity

Building Programmes. The FPO is engaged in the processing of vegetables, chili, tomato, ginger potato, cucumber, arbi as well as fruits such as apple, apricot, mango, rhododendron (Buransh)etc. The FPO received its own FSSAI License number: 22617026000012 from the Department of Food Safety, Government of Uttarakhand for processing of fruits and vegetables and the farmers from more than 50 SHGs formed in the villages itself are involved in processing activities and selling their produce to develop it as the tiny enterprises.

Keywords: *Farmer produce organization, Impact, Livelihood, Capacity building, processing*

Introduction

India is bestowed with a lot of potentials to produce a variety of organic products due to its diversified agro-climatic regions and this is very true in case of Dehradun district of Uttarakhand. It offers a variety of geographic locations suitable for cultivating various food crops such as: fruits, vegetables, food grains etc. Globally, organic farming is cultivated in an area of 57.80 million ha. However, India accounts for just 2.59% of the area. The organic farming is followed in India in almost all the regions with three states Madhya Pradesh (34.34%), Maharashtra (13.19%) and Rajasthan (11.67%) accounting for about 59% of the total organic area of the country. The cultivated land under certification has substantially increased from 1.74 lakh ha in the year 2005-06 to 1.78 million ha in the year 2017-188 (https://www.researchgate.net/publication/338209816_Impact_of_farmer_producer_organization_on_organic_chilli_production_in_Telangana_India) Agriculture and its allied sectors are the largest source of livelihoods in India. However, India still has many growing concerns. As the Indian economy has diversified and grown, agriculture's contribution to GDP has steadily declined from 1951 to 2011. (Source: http://www.fao.org/india/fao-in-india/india-at-a-glance/). One of the main reasons is due to the lack of competence in Indian farmers for selling the crop in remunerative prices in the market. Some of the critical constraints and bottlenecks are lack of infrastructure for storage, transportation, quality control, packaging, price risk management, cool chains, market led extension, and framework for promotion of contract farming (Gohain, 2018). Realizing these issues Government of India under the Department of Agriculture and Cooperation identified Farmer Producer Organization (FPO) as the most appropriate institutional form to mobilize farmers and build their capacity to collectively leverage their production and marketing strength(Ministry of Agriculture, 2013).For analyzing the potential, importance and contribution of FPO in the agriculture sector, the present study aims to discuss the role of

FPO in uplifting small and marginal farmers and its impact on the basis of the project undertaken by *Silgur Bizat SwayattSahkarita* FPO in Dehradun district of Uttarakhand.

Research Context

Sahiya village is located in Chakrata block, Dehradun district, Uttarakhand and serves as the gateway to Jaunsar-Bawar tribal region of the Dehradun district. Sahiya offers a tremendous opportunity for vegetable growers. This hilly area is bestowed with congenial agro-climatic conditions and has enormous potential for the production of off-season vegetables like, tomato, capsicum, green peas, beans, cabbage, cauliflower and cucumber, which are successfully grown in the season when these are not available in plains. There are number of farmers engaged in vegetable cultivation for their livelihood. However, due to unavailability of proper marketing systems, lack of processing facility and exploitation by middlemen, the farmers were not able to harness the full value of their produce. Hence, it was decided to process the fruits and vegetables available in the local area to reduce fruit and vegetable losses and to reap good profit from the produce. The intention behind this move was to improve farmers and their families' nutrition diet by allowing them to consume their own processed fruits and vegetables and to develop new value-added products. Another objective was to generate both rural and urban employment and to make the fruits and vegetables available during off-season which will directly lead to generate new sources of income for farmers/artisans.

Methodology

The vegetable growers who were interested in processing of fruits and vegetables were brought under one roof by an NGO named Society for Economic and Social Studies (SESS), Centre for Technology and Development, which works in Uttarakhand state. The farmers were motivated to form a Farmers' Producer Organization (FPO). Ultimately an FPO named Shilgur Bizat SwayatSahkaritaSakni (SBSS) came into existence and was registered under Society Registration Act, with registration no is T.C.09, dated 21st March 2016. The SBSSorganization is now processing fruits and vegetable into pulps/spices/dried products through a credit support fromNational Bank for Rural Development (NABARD). So far, six batches of trainings have been completed from Krishi Vigyan Kendra and Doon University, Dehradun along with exposure visits for FPO members. After the documentation and training work, the FPO started its production activities from April 2017 onwards. The processed products sell in the market under the brand name: Jaunsar Fruit Products. The FPO is engaged in the processing of vegetables, chili, tomato, ginger potato, cucumber, arbi as well as fruits such as apple, apricot, mango, rhododendron (Buransh)etc. The FPO received its own FSSAI License number: 22617026000012 from the Department of Food Safety, Government of Uttarakhand for processing of fruits and vegetables.

Project Interventions

One of the key aspects of this project is the post-harvest handling of Agro-horticultural produce, involving establishment of units for processing and value-addition. POPI-SESS has taken an innovative step by establishing a Common Facility Centre (CFC) in Mandi Campus of Village Sahiya by renovating an unused old building. Later, equipment &machines were installed in CFC for processing of fruits and vegetables. The CFC also facilitated and conducted anindustrial level training to the associated farmers, women, SHGs, etc. apart from the formation of farmers' groups, help with financial inclusion to micro-enterprises and thereafter promoting sales and marketing. This CFC is now managed & controlled by a Farmer Producer Organization – Silgur Bizat SwayatSahakarita (SBSS) by connecting farmers from 39 villages.

The objective of mobilization of SBSS was:

- To provide gainful employment in rural areas
- To help the rural people to acquire the basic skills which would enable them to be gainful employees in rural vocations
- To expose them to adopting a new technology to boost the production, both in agriculture and rural industries
- To arrest their tendencyto migrate from rural areas to urban areas in search of employment and livelihood to continue to live in object poverty as well as expose themselves to the hazards of the urban environment
- To help the rural families to emerge from poverty within a short period of time
- To provide a platform for other departments/NGOs to conduct development programmes

Discussion

Major activities of FPOs

1. **Utilizing Local Produce:** Local produce like Mango, Rhododendron, Lemon,Gooseberry, Ginger, Garlic, Cauliflower, Capsicum, etc. were processed and convertedinto various products like juice, jams, jellies, squashes, murrabas, candies and dehydrated products. Various legumes and lentils like kidney beans, toor dal, Amaranth,etc. were graded and packaged.

2. **Backward to Forward Linkages**: To ensure the continuous supply of raw material in CFC forits sustainability, 64 Ha of land was planted with Anar, Lemon, Amla in 5 villages namelyAlsi, Sakni, Kanbhua, Kakadi and Kimotha under JaunsarTribal Development project.Lemon and Anar have already started fruiting in some patches. Few crops are nowreaching CFC for value addition.
3. **Introducing Mechanical Processing:** It has been the first time in Sahiya region that food processing equipments& machines were introduced to local community for processing.These machines were procured through conversion with Integrated Livelihood Support Programme (ILSP) and Uttarakhand Council for Science & Technology (UCOST). Villagers were trained enough to handle machines like juicer, pulper, paste making machine,ginger peeling machine, spice grinders, sealing machine and many other useful instruments.
4. **Premier in bank loans for processing**: Silgur Bizat SwayatSahakarita, after getting appropriate training at CFC Sahiya, has got Rs 2,00,000 loan from Punjab National BankSahiya in April 2017 (Loan Account no: 3128008700000999) and started production &marketing of processed products under their own Brand "***Jaunsar Fruit Products***".
5. **Market linkages**: FPO is able to tie up with other FSSAI units in Dehradun like Future Food Products Selaqui, CTD Sahaspur besides participations in state level fairs such as Saras Mela- Dehradun, Career Mela- Kalsi and Saras Fair- Pragati Maidan New Delhi. For procurement of raw materials, board of members (BOD) and POPI Representatives met with two federations of ILSP in TyuniChakrata.

Farm Field Activities Performed by SBSS

Maize Sheller: One Maize Sheller (Thersher), designed by Vivekananda Parvatiya Krishi Anusandhan Kendra (VPKAK) in Almora, was provided to the SBSS- FPO under the TribalSub Plan. This has helped a lot in the reduction of women drudgery. A one dayorientationprogramme was organised by VPKAK at CFC Sahiya about its application and usage. The maizesheller removes the kernels in the gentlest possible way by rubbing the cobs againsteach other without any risk of harming the germination capacity of the sensitive maizeseed.

Chamomile Cultivation: Under the technical guidance of Doon University, 10 Farmerscultivated 11.5kg of Chamomile seed in 10 beeghaland.This land was previously used for potato cultivation. Although, the farmers were new to its cultivation, yet they generated goodyield (77kg) of chamomile flower. The flower dried in biomass drier and were sold in marketat a better rate then potato. This practice is being repeated from the last three years.

The impact of the FPO was measured under six parameters which are described as follows:

1. **Livelihood Related Impact:** Setting up a CFC in Sahiya with convergence with S&Tinstitutions/ NABARD TDP fund and providing financial inclusion from PNB Sahiya to SBSS FPO were the two benchmark for this program and they resulted in stable livelihoods, reduction in distressand migration, improved quality of life, increased capacity and participation of women in development process, creation of entrepreneurship and vibrant communityorganizations.

Activity	**No. of Beneficiary families**	**Income (Rs.) pre-development 2015**	**Income (Rs.) post-development 2015**	**Impact**
Promotion of SBSS through Processing of fruits & vegetables Under the brand "Jaunsar Fruit Products"	>200 Families	Mostly for own consumption turnover	90,410/- profit as per audit report 2017-18 Turnover: 1,39,5892,25,00/- profit as per report 2018-19 Turnover: 3,40,000/-	CFC in Sahiya has became a backbone for the whole Kalsi region

2. **Agricultural Development Related Impact:** Cultivation of Chamomile flower in 10 ha in Jaunsar region & introduction of hybrid varieties of vegetables such as coriander, maize,cucumber, karela, french beans have affected the cropping pattern leading to increase in the commercial crop production and their productivity. Utilization of 2ndgrade of fruits in processing also helps farmers in its marketing.
3. **Migration Related Impact:** During the project period, there was no migration in theproject area. This can be attributed to the activities carried out in the CFC for the income of the farmers. New and better livelihood opportunities are available to the villagers in their villages itself. More & more community-based organizations (>50 SHGs)were formed in the villages itself to develop the tiny enterprises.
4. **Quality of Life Related Impact:** The traditional method of pickling, jam making, squash making, sun drying vegetables and spices in Jaunsar region was a timeconsuming process with a high probability of pollutant contamination and/orvulnerability to rodents, pests, insects, and other such threats, rendering thequality of the produce unsafe for human consumption. Innovative approach was to introduce improved processing techniques that wereabsent in the whole region between Kalsi and Chakrata. The local farmers/SHGs have been motivated to form a Farmer Producer Group-

"Silgur Bizat SwayatSahakarita" with the support of NABARD. This group is undertaking processingactivities in this CFC with a credit support from the bank.The production activities started April 2017onwards. The processed F&V products sellsin the market through the brand name of ***Jaunsar Fruit Products***.

5. **Environmental Impact:** Food preserved permanently by addition of preservatives likesugar, salt and vinegar and chemicals (Sodium Benzoate, Potassium Meta Bisulphate,Sulphur Dioxide, Citric Acid) in sufficient concentration to prevent the growth ofmicroorganisms either by osmosis or by poison or by both. Under FSSAI act, all thesechemicals are made from food grade ingredients and thus environment friendly.
6. **Impact on Women Development with specific reference to gender issues:**The women development related impact on account of their participation in the projectis examined in detail with reference to specific gender issues. The issues related to theirsocial importance, responsibility, roles in decision making, participation in family welfareas well as the developmentoftheirowncapabilityandeconomicindependenceareassessed. The chief strategy for women development is through development of morethan 50 new Self Help Groups (SHGs). This workforce has been engaged for bulk semiprocessing, storing and packaging of raw material. Its further linkages with SBSS-FPO hasensured women empowerment, food security, healthcare, education for children,hence, prevented migration.
7. **Convergence related impact**: The activities under NABARD's initiatives are being undertaken and are expected to improve productivity and quality of farm and plantation activities. But for the sustainable utilization of this CFC, training infrastructures has been created and thus the tribes residing in surrounding villages could be added for the promotion of livelihoods. Hence, UCOST & ILSP support has been very useful in establishment of TRC in Jaunsar region itself.
8. **Impact of Capacity Building Programmes:** The training which was provided at CFC Sahiya, had the additional advantage of being conducted in the environment ofrunning enterprises, thus providing trainees with on-the-job practical training andexposure to rural enterprise management at field-operation level.The CFC was inaugurated in October 2016 and up till now,10 batches, covering45 villages and 354 beneficiaries trained so far in different aspects of foodprocessing. SESS is also utilizing the funds from Tribal Development Project,Integrated Livelihood Support Project and SARTHI Project through DST- GOI forthese beneficiaries apart from UCOST. Details are:

Details of Training Programmeson F&V Processing			
Date	Days	Participants	Venue
23 Oct- 5 Nov 2016 7	7	90	Village Level
6-8 Nov 2016	2	27	CFC Level
16-21 Dec 2016	5	25	
23-28 Dec 2016	5	30	
2-6 Jan 2017	5	25	
16-19 Feb 2017	4	20	
15-17 March 2017	1	42	
14 Sept 2017	1	37	
26 Sept 2017	1	31	
10-11 Nov 2017	2	27	

Outline of the Training Curriculum(at initial phases)	
Principal of Food Processing: Bulk preservation and storage of temperate fruits Operation system of equipments Storage of juices & pulps Conversion of pulp into squashes Product formula of pickle Preparation and principle of jam, jelly making Product design of fruits &vegetables Product formula for sauces & chutneys Quality control procedures FSSAI- its content andregularity process Spice processing Vegetable paste Dehydration technique Packing and quality control Cost benefit analysis Marketing	
Product developed so far:	
Pickle	Amla, Chili, Mango, Lemon, Arbi
Squash	Mango, Lemon, Burans
Jam	Apple, Mango, Mix Fruit
Chutney	Apple, Amla, Lemon
Candy	Amla

Conclusion

Agriculture has always been a lifeline of the Indian economy, providing livelihood to millions of farmers.However high production costs, and low access to credit, as well as poor market linkages have always remained the barriers to the sector's growth. Farmers Produce Organizations (FPO) are the possible solutions to give farmers better bargaining power and create a more apparent agri-market, to help the new generation of farming groups.

SilgurBizatSwayattSahkarita FPO in Dehradun district of Uttarakhand started its work with different constraints in its establishment and development, viz. finance, human resource, local competition, knowledge, and technical skills, etc. It created an excellent platform to implement innovative and productive activities in a mutually benefitting situation. All the efforts put in by the people, resulted in a symbiotic growth, both in financial, motivational and in the overall development of the farming community.

References

Gohain, N. (2018). An Analysis of Problems and Constraints Faced by Farmers in Marketing of Agricultural Produce in Punjab. Economic Affairs, 63(3):671–678. https://doi.org/10.30954/0424-2513.3.2018.11

Manaswi, B.H. (2020) Impact of farmer producer organization on organic chilli production in Telangana, India ndian Journal of Traditional Knowledge. 19(1):33-43.

Ministry of Agriculture, Government of India. (2013). Policy & Process Guidelines for Farmer Producer. SFAC, 95.

28

Role of Government and NGO's for the Challenges Being Faced Small Women Entrepreneurship in one Specific Territory

Rashmi Chaudhary and *Pooja Prakash Srivastava

IBS, ICFAI University, Dehradun, Uttarakhand

Abstract

The study aims to stress on problems being faced by female entrepreneurs in India who have been operating in different fields of business and thriving hard to make a position in the business world. On one hand with the emergence of e-commerce, economic development and government support, female entrepreneurship has flourished over the past few decades. However, on the other hand, many women entrepreneurs are facing the typical traditional barriers such as gender discrimination, lack of support from family members, negative approach of towards businesswomen. Women have always been considered as a non-serious businessperson. The paper thoroughly reviews the extant literature case note and contemplates upon the challenges being faced the small women entrepreneurs in creating new Start-up's.

Keywords: *Female Entrepreneurs, Problems of Women Entrepreneurs, Gender Biasness, Start-up's, Growth strategy*

Introduction

India has been advancing excessively in terms of workforce engagement as a lot of women have started seeking space in the business arena. The new age has given way to a new trend of entrepreneurshipwhich is flourishing without the quantification of gender. Needless to say, India has been dealing with a very bad history of patriarchal class arrogance and work division that cannot fully provide all the freedom to women who want to prosper in their professional lives.

The female entrepreneurs have to undergo a lot of obstacles right from the initiation of the idea till the plan operates. Sometimes, the existence as a women itself comes with a baggage of responsibilities. Indian women are bound to fulfill their responsibilities towards their family, community and work, balancing everything equally and in an efficient manner.

The conventionalism, ethnicity, social and cultural principles, parenthood, the tag of feeblest sex and insecurity curb the confidence of women in comparison to their male counterparts. Even though there have been considerable changes found in the women dwelling in urban areas, the rural women still have numerous hindrances blocking their path. The perceptive of the opponent sex is irrational in rural areas due to the lack of education and it becomes hard for the women to resist men in such cases.

The Indian customary draws a thin line between the capability and responsibility allotted to different gendersMohsen B., and Zouari S. (2014). It has even threatened the employment chances of females in the traditional and eccentric sphere of fiscal exercises. The present scenario has not broken the undeterred confidence of women who sought to make a career in business, since the last decade. With the alteration in technology, globalization and rivalry accentuation has made the operation of business more multifaceted and vibrant.

Female entrepreneurs are seen to be excessively inspired and self- directed. However, as the competition rises, entrepreneurs should have the ability to shift strategies rapidly and habituallyBrahmi, M. and Laadjal, A., (2015).The inner mental litheness, visualizationfor an elongated planning and the capacity to endurevagueness and alterations are an advantage for every business organization and also their skill toconfer in promoting the merchandise to potentialshareholders, dealers and patrons.

There are certain emotional and social barriers that ties women in its burden that can affect her work performance.Indian women are seen to run a family right after they are married. The entire stress of taking crucial decisions in the house to the household chores is on their shoulders. Women are also tagged to be the most vulnerable beings in terms of emotions and have a lot of responsibility towards their family and children. Husbands and in-laws with orthodox mindset want to be in house to attend their home first which in turn can complicate her professional life and cause problems in running their venturesproductively.

The most noteworthy encumbrance that women commonly face while attaining success, is the male ego. Irrespective of all the sermons and orders that supports gender equality, our society still follows the patriarchal form and looks down upon women who want to rise and become equal. Females are still required to take the permission of the head of the family before starting anything of their own. Chauvinism in husbands and other male members of the family can create issues for working women. Priorly, entrepreneurship has been the space for male dominants which was later being occupied by women. Therefore, male members still create objections against their female counterparts.

India has been lacking far behind in terms of girls education. More than half of the girls in rural India are uneducated due to several reasons. The enrollment of girls in different educational institutions is less as compared to boys and the dropout rate is also high as some of them are married at a very tender age and some of them are forced to discontinue schooling due financial issues of the family or other social concern. The behaviour can cause difficulty in making prudent changes in the business later on. Women with insufficient knowledge cannot implement necessary technological, production related and promotional changes. These factors can be a challenging threat for the success of the business.

Women are often tabooed for their incapability to achieve successfully and are stereotyped for being less intelligent and dumb in corporate affairs. Businesswomen have to struggle a lot in order to mange finances for a smooth functioning of their enterprise. Many financial institutions, bankers and investors do not readily agree to invest or provide loan to a women run venture because of less credit worthiness and their perceived chances of business breakdown. Women also face challenges in terms of obstacle of finances in raw materials, work-in-progress processed goods and unrecovered payment from consumers.

Another crucial peril that knocks down the accomplishment of women is the stiff competition that women face from all the elements around them. The lack of knowledge of technology in the manufacturing procedure that is adopted by women can impose a severe hazard in the growth of business. The market competition is soaring as male marketers have better strategies of survival while the female marketers are having a tough time in finding their relevance of existence in the prearranged sector as men running business have more experience and aptitude in administering an organization.

Several determinants that include unproductive supervision can add to the increased cost of manufacturing that stands as a hurdle in the career escalation of women entrepreneurs. The basic drawbacks in women industrialists are the technological ignorance due to the unawareness or slow adoption to the transforming expertise can be a vital reason for the elevated cost of invention.

Women areoften projected as weak, introvert and gentle in nature. Sometimes it becomes hard for women to put up with the market weather while operating a business. Inadequate education, guidance and lack of monetary assistance from the sources also decrease the capacity to stand the jeopardy that are attached with entrepreneurial activities.

India has also been a country that lacks in its ability to provide safety to women. Therefore, the mobility of women becomes restricted restraining all the possible opportunities that are needed to flourish their business. Other grounds for immobility are due to the conventional morals of the Indian custom that confine women from a lot of work. A single women working till the late hour at night is often looked suspiciously by the society. Having male workforce who have bad intentioncan make the work ambience uncomfortable for women.

Deficiency in the entrepreneurial propensity is also a subject of apprehension for businesswomen. It is even alleged that women don't have the business bent of intelligence.It is as noticed that in spite of all the instruction and training schemes on entrepreneurship women marketers becomes unsuccessful in fighting all odds related to managerialfunctioning.

Supervision has become a generalized field that can only performed by proficient experts. The female entrepreneurs are not found to be competent enough in administrative job like development, management, scheming, harmonizing, recruiting, direct, inspiring, etc. of an organization. Consequently, fewer and partial decision-making capacity of women has also been a crisis for them to operate a venture productively.

Gratifying the officially permitted paperwork that is needed for the succession of an organization is a complex chore for the women entrepreneur due to the occurrence of fraudulent exercise in the administration sectors and bureaucratic postponement for diversecertification related to electricity, water and shed allocation. In the given condition, the women who run the business find it intricate to focus on an evenrunning of the endeavor.

The women in business have also fallen prey to the middle men involved in marketing. As a huge part of the business is based on its promotional activities, it is not possible for women to roam around for marketing, distribution and fundcompilation; therefore they have middle men to do the above mentioned tasks. Thesepeople are seen to abuse them while pretending to help.They attached an excessively high amount of commission charges for their services which leads to decrease in profit.

Businesswomen are entitled with the lack of self-assurance which is an omni potent aspect in the smooth functioning of business. They have to struggle harder in order to maintain poise between work and family. It is their great sacrificing power that it's seen, many a times that women giving up on their dreams for the sake of their family.

Review of Literature

Sumathi and Gunasundari (2016)state that women entrepreneurs are defined as acluster of women who commence, manage and run a business venture. The administration of the country has stated that the female industrialists are the ones who own and controlan enterprise and have atleast more than half, specifically 51% of monetary interest and give atleast 51% of the service in the industry to women.

The female entrepreneurs have similar functions to that of male entrepreneurs. The services involved in the operation of their venture is beginning a new enterprise, assume risk factors, overture of fresh advancements, organization of management and power of the trade and catering effectualheadshipin all spheres of business.

Agarwal (2018) mentions that female industrialists are denoted to have included about 1/3 of the universal entrepreneurs.The Indian women entrepreneurs are said to have brought a tremendous changes in terms of development in the private and public arenas.The result of globalization and its succeeding resources are getting included as spiritually just like the male proportion in the country. The initiation of the entrepreneurial programmes for women in India has gain acceptance and recognition all over the world. Even though the market has expanded and e-commerce has advanced enormously, women are succumbed with a variety of cons in order to maintain the market pace.

The contribution of women in the Indian workforce is very low in comparison to the universal parameter.The most prevalent and respected job is the job of white-collars that is discontinued by women due to family or societal pressure. Some of them choose to venture on their own business. It is even seen that women have been successfully operating home based business like baking, weaving, printing, etc that gives them the liberty to manage home and work together and help them be under societal customs.

However, the contemporary women have gone beyond the boundaries and explored fields that are associated with computer know-how, internet and improved communication apparatus. Many women are not the bread earners of the family but have started entrepreneurial business in order to support their family and become independent. They want to make the skillful use of their education and invest their time into something fruitful after the completion of their daily household chore.

Homemakers turned businesswomen can face a lot of challenges related to family and the society. Therefore, they love to accept occupations that can allow them to operate from home (Bose, A., 2019).

Kumar and Balamurugan (2006) focused on the stigmas that are associated with women as they are considered to be the home makers with limited knowledge and no relation with the economy and trade.There have been certain positive noticeable changes in the perspectives of people.The modern world has opened more opportunitiesfor women and has allowed them to take up entrepreneurial activities in medium to small scales.

It is noted that women play a very crucial function in making the backbone of the nation's economy. The study concludes that the craze of entrepreneurs is high among women who have completed their post graduation.

Taramol (2015) illustrates that fiscal advancement is the consequence of assistance from the workforce including both male and female. The position of women is transforming through the universe and they are no more restrained in their houses but have enormous responsibilities to authorize their prospective.

Right after India got freedom, the policies and schemes about the monetary development was restructured giving more allowance to women on a national basis. Progress in the field of micro ventures can assist in making spontaneous servicechances to innumerous women at a very low investment.The author thinks that the micro enterprise scan be a great option of employment for people with poor financial condition in turn elevating their profits and standard of living.

It is believed that when a business is managed by women, there are more rational changes and enviableupshots that are observed.India is one of those third world countries that are progressing in terms of education and service sectors.In spite of 40 years of advancement, it has to go a long way in order to compete with the first world countries in the world.

The objective looks far but the amount of time that is needed to accomplish it can be curtailed by speeding up the progressing its speed and also by marketing of industrial and business enterprise that are formed by the effort and proposal of the business class.It was during the late fifties when the trend of entrepreneurship actually came into existence and it slowly gain popularity in many other parts of the world.

This has proven to be the best source of income if one posses the necessary skills. Some relevant economists have explained the function of entrepreneurs in their own words. Schumpeter (1949)thinks that prologuemodernism is one of the most noteworthyinputs of industrialist in financialexpansion. Parsons and Smelser defines that entrepreneur is one of the two most vital situations that are needed for a monetary appraisal.Harbinson (1987) believes that entrepreneurs are the key aspects of modernization.

The recent research is a step to scrutinize the bottlenecks that are faced by female capitalists of the urban area of Rohtak district and the social and economic backdrop of themand also their struggle story from the initiation of their venture till success (Vijender, 2019).

Abdul, K. Gunasagaran, N. A. Kamarudin, M. Rahman, J. Uddin and S. Rahaman, S. (2017) in their paper scrutinizes the important aspects and obstacles of women entrepreneurs in Klang Valley, Malaysia.The research exposed that 85% of the women have the knowledge of business before associating with it and 85% of them also deal with an enterprise that comprise of atleast on of their family members.

On the basis of ranking given Pearson correlation coefficient,aptitude is the most prominent thing needed to run a business which is again backed by business policies, ambience, outlook and utilization of technology, chronologically. The exploration also excavated some issues that are faced by business women during the function of the business. The analysis of the research can be a very helpful document for the administration in order provides appropriateeducation or guidance to Malaysian women entrepreneurs.

Mauchi, M. Mutengezanwa, and D. Damiyano (2014) reported that the women entrepreneurs have worked incredibly for the economic development of the country, but their recognition has remained passive. It is a clear indication of all the gender related complexities that women have to face in the male dominated society. The prime most objectives of the research were to unleash the challenges of the female industrialists in Mashonaland Central Province of Zimbabwe.

The exploration suggests that middle aged women entrepreneurs who are engaged is functioning as sole proprietors or are in a partnership pact. However, most of them don't have a proper registration. It is concluded that women who run business are often challenged with difficulties pertaining to funds, family, and lack of adequate knowledge about the latest expertise and also restraining rules of the society. Sourcing raw materials markets were cited as the least challenges for women entrepreneurs. Sourcing raw materials is denoted as the least difficulty a women entrepreneur can face.

Mohamad and M.S. Bakar (2017) invisioned about Malaysia being the most progressed country in 2020. One of the biggest reasons for such advancements is the encouragement that they provide to fellow women entrepreneurs. They empower the women workforce in order to escalate the poverty of the country. The current administrative power in boosting the morale of businesswomen and have also denoted to be the catalyst of the economical transformation in the country.

Many women choose to contribute to their family by working on a full time and part time basis. Malaysian women workers associated with business do face a lot of threats while operating business. The research states that in order to overcome the following challenges of the women, it needs to be addressed. Projection of their weakness can make them stronger and will help them face challenges confidently. The Malaysian women entrepreneurs have to be empowered and motivated and they can enhance the economic position of the country to a great extent.

For Srivastava (2018)mentions that women have an innate capability to manage things incredibly. They have an inborn capacity to synchronize, ideate, execute, supervise and align activities as they have proved to be the only one to be capable of being the perfect home maker. Women have started acknowledging their power with their exposure to education and knowledge and have even fought to break all the bondage of societal specifications.

There has also been an affirmative response from the rural areas of India along with the urban areas.The rural areas have started gaining the participation of women in entrepreneurship for the elevation of local wealth.Their indulgence has boosted their confidence and has made them realize the significance of being an active hand in the development of the nation.

By the different modes of contributions they have understood the relevance of financial development and about independence. Their journey from dependence to being independent has amazed them and has increased their respect towards work. Apart from all the challenges and restrictions, it is their determinedstrength that they have designed their own fate and have attained triumph.

Conclusion

Women have always been considered to be the most susceptiblegender in the Indian society. The burden of the orthodox community and the vulnerability that is attached with the origin of women has been the biggest hurdle for her to achieve victory in any principle. However, with the passage of time, India has adopted a new perspective to not judge women as only the home maker but a great investor in the country's economy.

Previously, thinking about anything else apart from the family was considered a sin for all the married women in the country. But with the advent of proper education and the modern gospel, women here have improved drastically in all spheres of operations. Needless to say, a successful women with a smooth functioning of business, is not at all a digestive element for the male dominated society in many cases.

The research was done in order to explore various threats that block the chances of women to attain victory in the business arena. An Indian woman before entering into a venture has to confront the arrogance of her male counterparts. The conventionalism has taken a toll on the modern era with the revised strategies which states that women play a very considerable role in elevating the financial structure of the economy but raising funds for women led enterprises is a tough task as they hardly have any collateral security to claimdebt.

Often, their credit worthiness stands as a hurdle to a smooth functioning of business. Another important disadvantage is the cut-throat market competition that women receive from men in order to flourish further. The family ties and emotional instinct works strong for a women and she cannot wash her hands off from the responsibilities. Traditional believes and cultures followed in Indian households often restrict women entrepreneurs to follow their dream. The conventionalism also curbs the mobility of women.

As women travelling for their professional need is suspiciously looked down by people.It is also noted that women abroad are given more relevance in terms of professionalism and are considered to be an equal to men. But Indian women first have to face the dominance of their male equivalent in order to survive in the business atmosphere. The prime most downside of less participation of women in business is the lack of education.

Girls have a lower rate of enrollment in schools and a much higher rate of dropouts due to financial and cultural issues. It leads to the lack of expertise and technical knowledge which can act as the biggest hindrance in the functioning of their enterprise.Also, there have been a lot of stereotypes that are attached with the image of women as they are measured to be weak, shy, mild and unintelligent.

Though women nowadays have got enough liberty to build their own career, there are certain loopholes that need to be covered. Men should understand that running a house is not the sole job of their wives while women should respect and stand for themselves beyond societal clause. An effective strategy developed by the government to train and educate women through several social programmes can help in eradicating gender difference in entrepreneurship. A broader perspective and a warmer attitude from the society along with soulful assistance can help women fight with all hurdles and achieve even more in the corporate world.

References

Abdul, M., Gunasagaran, K., Kamarudin, N. A., Rahman, M., Uddin, J., and Rahaman, S. (2017). Critical success factors and challenges of women entrepreneurs in Klang Valley, Malaysia. Journal of Advanced Research in Business and Management Studies, Vol. 6(1), pp. 12-23.

Agarwal, M. (2018). Challenges Faced by Women Entrepreneur in India. International Conference on Social Sciences, Humanities and Education, Vol. 18 (2), pp.41-56.

Bose, A. (2019). A Study of Women Entrepreneurs in India. International Journal of Management (IJM), Vol. 10(4), pp. 132-140.

Brahmi, M. and Laadjal, A., (2015). The Strategic Choices of Small Medium-Sized Enterprises Integration: Evidence from Specific Economic Territory, AUDŒ, Acta Universitatis Danubius. Œconomica,Vol. 11(2), pp.254-271. Retrieved from:http://www.journals.univ-danubius.ro/index.php/oeconomica/article/view/2731/2399https://www.ceeol.com/search/article-detail?id=521810

Kumar, M. S. and Balamurugan, S. (2006). An Empirical Analysis on Stress Level of Women Entrepreneurs and the Factors Responsible for Stress,journalresearch management and practices, Vol. 3(1), pp.56-64.

Mauchi, F. N., Mutengezanwa, M., andDamiyano, D. (2014). Challenges faced by women entrepreneurs: A case study of Mashonaland Central Province. International Journal of Development and Sustainability, Vol. 8(8), pp. 466-480.

Mohamad, M., andBakar, M. S. (2017). Malay Women Entrepreneurial Success: Challenges and Barriers. International Journal of Accounting, Finance and Business, Vol. 2(5), pp. 78-84.

Mohsen B., and Zouari S. (2014). The Employment Policy in the Mining Industry: A Critical Study on Gender Discrimination, Journal of Asian Business Strategy, 2014, vol. 4, issue 8, 98-107,http://www.aessweb.com/pdf-files/1-142-4(8)2014-JABS-98-107.pdf

Srivastava. A. (2018). Contribution of Rural Women Entrepreneurs in Nation Building. International Journal of Social Science and Economic Research, Vol. 3(9), pp. 68-71.

Sumathi, V. and Gunasundari, K. (2016). Problems Faced by the Women Entrepreneursin India. International Journal of Management Research & Review, Vol. 6(2), pp. 348-354.

Taramol, K.G. (2015). Scio-Economic changes of Women Micro Entrepreneurs in Kerala. Research Journal of Economics and Business Studies, Vol. 4(9), pp.152-171.

Vijender. (2019). Women entrepreneurs: issue and challenge. The research journal of social sciences, Vol. 10(4), pp. 324-337.

29

Sericulture Development in India with Special Reference to Public-private Partnership

Saniya Tyagi and Ram Keval

Department of Entomology and Agricultural Zoology, Institute of Agricultural Sciences, Banaras Hindu University, Varanasi, Uttar Pradesh

Abstract

India is the second biggest producer of silk in the world. Sericulture industry has occupied a prominent place in the industrial development with low investments, gestation period and high employment potential and benefit-cost ratio in rural and semi-urban areas. This chapter aims to highlight the functions and role of the various infrastructure of the public (central and state government) and private agencies providing various services for the sericulture industry. There are also some recommendations given on the reorganization of this infrastructure for its improved effectiveness and cost-efficiency by highlighting the development of sericulture sector through public-private partnership in the country.

Keywords: Sericulture, Privatization, Silk, Employment, Public-private partnership (PPP)

Introduction

Silk also known as the "Golden Fibre" is admired all over the world for its lustre, strength and smoothness. Sericulture industry is an ecofriendly agro-based activity of mulberry cultivation, silkworm rearing and production. It is a labour intensive, rural cottage industry providing subsidiary employment and supplementing the income of semi-literate and semi-skilled poor workers throughout the year [Chanotra*et al.*, 2019]. The term 'sericulture' includes economic activities which receive poor workers in the value addition at each stage like (i) raising plants for feeding the silk-worms (in the form of the mulberry plantation); (ii) production of disease-free silkworm layings (dfls), eggs, or seeds; (iii) indoor rearing of silkworms till the cocoon stage which produces silk (about 30-day egg-to-cocoon cycle), and (iv) sale of silk cocoons produced.

Silk industry can be broadly classified into; (a) sericulture, (b) post-cocoon technology – reeling (yarn production), spinning and twisting of silk yarn from the cocoons; and (c) weaving, fabric production, printing and dyeing, finishing, garment designing, marketing etc. These activities also provide the auxiliary enterprises of by-product utilization and machine manufacture including spun silk yarn manufacture and pupae oil extraction, thus creating livelihood for the small and marginal farmers and labour force.

Asia is the key producer of silk in the world producing over 98.5 per cent of the global output. There are more than 40 silk producing countries, but the majority of its production comes from China and India, followed by Brazil, Uzbekistan, Thailand, Vietnam, Korea and Japan. Unlike temperate countries, India has a distinct advantage of practising sericulture all around the year, as a result of its tropical climate, yielding a stream of about 4-6 crops (Savithri and Sujathamma, 2019).

Sericulture Sector in India

India is at present is the world's second largest producer and the largest consumer of raw silk and silk fabrics. In 2019, The Indian sericulture market was worth 266 Billion rupees. The total raw silk production in India during 2018-19 amounted to around 35,468 MT (CSB report, 2020). India is expected to become self-reliant in the production of silk anytime soon.

Types of Silks in India

India is the home of the vast variety of silk-producing insects. India achieved a unique distinction of the occurrence of all the five major types of silk of commercial importance obtained from different species of silkworms. These varieties of natural silks are Mulberry, Tropical Tasar, Oak Tasar, Eri and Muga. Except for mulberry, other non-mulberry varieties of silks are generally known as vanya silks. The majority of the commercially produced silk in the world is mulberry silk which comes from the silkworm, *Bombyx mori* L. It feeds solely on the leaves of the mulberry (Morus sp.) plant and is domesticated. Among the four varieties, Mulberry accounted for 71.45%, Tasar- 8.4%, Eri- 19.48% and Muga- 0.66% of the total raw silk production in the fiscal year 2019 (CSB report, 2020).

Table 1: Commercially Exploited Silkworm species of the World

Common Name	Scientific Name	Origin
Mulberry Silkworm	*Bombyx mori*	China
Oak Tasar Silkworm	*Antheraea yamamai*	Japan
Oak Tasar Silkworm	*Antheraea pernyi*	China
Oak Tasar Silkworm	*Antheraea compta*	India
Oak Tasar Silkworm	*Antheraea frithi*	India
Oak Tasar Silkworm	*Antheraea proylei*	India
Tropical Tasar Silkworm	*Antheraea mylitta*	India
Muga Silkworm	*Antheraea assama*	India
Eri Silkworm	*Philosamiaricini*	India

Source: Central Silk Board, India (http://www.csb.gov.in/silk-sericulture/silk/)

Organizational Infrastructureof the Sericulture in India

Public infrastructure for the silk industry in India exists at two levels: a central agency, that is, the Central Silk Board (CSB), which comes under the central Ministry of Textiles; and the state level Directorates of Sericulture (Census, 2012). CSB is the apex body for the overall development of sericulture and silk industry in India and is responsible for coordinating and implementing collaborative research and development activities with national-level institutions, state sericulture research and development institutes, universities, CSIR/ICAR/IARI/IITs/IIMs and other private institutions. Research and development (R&D) achievements of CSB like technologies patented and commercialized are aiming towards attaining higher quality and productivity levels of Indian silk.

The State Government Infrastructure

Out of the 26 silk-producing and consuming states, only 17 have a separate department or directorate of sericulture. The leading silk producing states e.g. Karnataka and Andhra Pradesh are the only ones having their own R&D Institutes for sericulture. The majority of the state governments' infrastructure includes:

- Silkworm Seed Production Centres
- Extension centres for the supply of input and new technologies transfer
- Reeling training and farmer training schools
- Market infrastructure for the sale of cocoon and raw silk.

For the development of sericulture in the country, the Government of India has introduced numerous schemes and programmes from time to time. During 2017-2020, the government has allocated ₹2161.68 crores to its Central

Sector Scheme 'Silk Samagra' for three years (PIB, 2019). It is an integrated scheme for the development of silk industry with the four components; (a) Research & Development, training, transfer of technology and I.T. Initiatives, (b) Seed organization, (c) Coordination and market development, (d) Quality certification systems, export, brand promotion & technology up-gradation. This integrated scheme will help in increasing the production of Bivoltine silk (high quality) by 62 per cent by 2020.

Privatization in Sericulture

India has a huge unexploited potential for silk production due to various factors. Recently silk imports from China have been badly hit on account of Coronavirus. This has led to an increased cocoon production in India to meet the growing demand for indigenous silk. It has been confirmed in many socio-economic studies that the benefit-cost ratio in sericulture is highest among comparable agricultural crops (Bharathi, 2016).Therefore privatization of this industry can be seen as a vital dimension to exploit this unforeseen potential.

As per now, the role of privatizationin the sericulture industry in India is very limited and only restricted to Silkworm Seed Production Centres where the production of disease-free silkworm layings, eggs, or seeds takes place. Silkworm seed production is a specialized commercial activity mainly undertaken by CSB and the state directorates and by private licensed seed producers of four traditional states where a substantial proportion (50-60%) of silkworm seed production is being carried out (Mathur *et al.*, 2013). These producers also extend credit and quality control facilities to the people involved in sericulture. However, the balance demand is met by commercial silkworm seed production centres under the central and state government agencies. But these agencies turned out to be inherently financially unviable due to bureaucratic controls and high establishment costs. Therefore production and some infrastructural activities must be transferred to private entrepreneurs through suitable incentive schemes and contractual arrangements, to maintain quality and production levels etc. Efforts must be made to develop suitable partnerships with the NGOs, private and cooperative sectors for rendering alternative extension services. Sericulture cooperative societies with or without financing by the government must be encouraged and trained for commercial operation of nurseries, providing extension to their members. To give them push private commercial silkworm seed and disinfectant manufacturers and NGOs may be provided with incentives and partial financing for offering various products and services.

Sericulture and the Role of Public-Private Partnership

Public-private partnership (PPP) is a key tool for the rapid economic development of any country for the creation or management of infrastructure for the public purpose for a specified period, on commercial terms. All privatization efforts report more accountability, flexibility in production and delivery of products than public services. As PPP facilitates greater efficiency, technological advances and innovation in the production it promotes cost-effective development. Various other sectors of the Indian economy (roads, water, and ports) are already working in PPP model, it is estimated that if this model will be introduced in sericulture industry, it will be benefitted in terms of technology, finance, and overall efficiency which will reduce the cost of operation and thus leads to increase in the total production and hence, exports of silk.

One of the examples of public-private partnerships in the sericulture area isthe National Research Development Corporation (NRDC) which has been appointed as the nodal agency for commercializing the technical knowhow in India and abroad, has made a licensing agreement with Rainbow Agrilife India, Andhra Pradesh for commercialization of "NEMHARI-A Plant-based formulation for Management of Mulberry Root-Knot Disease". NRDC has ambitious plans to promote this technology to all sericulture regions in the world through a network of dealers. It is also taking this forward through different government support schemes of the Department of Sericulture and other allied Departments in various states.

According to an estimate, the sericulture industry would generate employment reaching to 100 lakh people by end of March 2020 in rural and semi-urban areas in the country (CSB report, 2019). This would demand strategic promotion of the sericulture as a poverty alleviation programme as this industry belongs to the small and marginal farmers. Sericulture has both forward linkages (twisting, weaving and dying) and backward linkages (mulberry cultivation, rearing of cocoon and reeling, twisting, weaving of silk) which provides employment mostly for women and economically weaker sections of society, thus it is crucial to develop this sector to reduce rural-urban imbalance in one side and reduction of poverty on the other side in the state. To sustain the private sector and motivate unemployed youth to the post cocoon sector govt has taken initiatives like setting up of cocoon banks, providing working capital to reelers for purchase of cocoons, incentives on the production of silk yarn, setting up of exchanges etc.

Sericulture industry in general faces countless problems like the high cost of production and low productivity,absence of a proper market, lack of transport facilities, poor information on market trend, absence of storage facilities, lack of finance, lack of technology penetration, prevalent rural nature, houses without proper ventilation etc. Crop instability, fluctuations in the cocoon price, and insufficient technical guidance from the extension workers were the other constraints expressed especially by the large-scale sericulturists (Kumaresan, 2008).Therefore PPP can come out as a potential solution to tackle these problems and can bring down the cost of raw silk by promoting mechanization and rationalization in the mulberry cultivation, silkworm rearing and silk reeling. PPP will also improve the financial viability of reeling activity through efficient utilization of by-products and proper and timely marketing facilities. Joint efforts of government and private sector would boost this industry (Bhat, 2014).

Despite several pieces of evidence from other sectors where PPP leads to increasing investment rate and compound growth, sericulture is lacking behind due to low investment in terms of land under mulberry cultivation and infected seed, lack of storage facilities etc. supplied by the government. If both the sectors work together in sericulture industry, this sector will definitely grow both horizontally (base will increase) and vertically (increasing productivity, efficiency in terms of cost of production). Therefore, this holistic approach will help in the reduction of poverty, rural-urban imbalance and increase employment, which is currently a big challenge for the government.

References

Bharathi, D. (2016) Sericulture Industry in India-A Source of Employment Generation. International Journal of Advanced Engineering Research and Science, 3(10):236881.

Bhat, T.A. (2014) An analysis of public private partnership in sericulture in Jammu and Kashmir state (India). Journal of Economics and Sustainable Development,5(11):121-6.

Census of India. (2012) Available on the Internet: < http://www. censusindia.gov.in/2011-prov-results/data_files/india/Final_ PPT_2011_chapter3.pdf>.

Chanotra, S., Bali, K., Bali, R.K. (2019) Sericulture: An opportunity for the upliftment of rural livelihood. Journal of Entomology and Zoology Studies, 7(6):1100-1103.

Functioning of central silk board & Performance of Indian silk industry. https://www.google.com/url?sa=t&rct=j&q=&esrc=s&source = web&cd=&ved=2ahUKEwikyo_jtcbtAhVg4nMBHa4wClYQFjAAegQIAhAC&url=http%3A%2F%2Ftexmin.nic.in%2Fsites%2Fdefault%2Ffiles%2FNote%2520on%2520Seri%2520CSB%25202019%252020%25203rd%2520qtr_0.pdf&usg=AOvVaw3cuhvCy39B K1a7C3iVISR9, Jan 2020.

Integrated Scheme for Development of Silk Industry. https://pib.gov.in/Pressreleaseshare.aspx?PRID=1580493, July 2019.

Kumaresan, P., Devi, R.G., Rajadurai, S., Selvaraju, N.G., Jayaram, H. (2008) Performance of Large Scale Farming in Sericulture–An Economic Analysis. Indian Journal of Agricultural Economics, 63:902-2016-67978.

Mathur, A.K., Roy, S., Mitra, J. (2013) Toward Sustainable Development Of Economic Sub-Sectors: Case Of Indian Sericulture. Journal of Security & Sustainability, 3(2).

Savithri, G., Sujathamma, P. (2013) Entrepreneurial opportunities in sericulture Industry. International Journal of Engineering, Business and Enterprise Applications, 52-56

30

Skill Development and Empowerment of Women in Food Industry: NGOs Unlocking their Potential!

Shatakshi Semwal[1] and Lalita Vatta[2]

[1]IUCTE, Department of Education, Maharaja Sayajirao University Baroda, Gujarat
[2]Mahila Maha Vidyalaya, Banaras Hindu University, Varanasi Uttar Pradesh

Introduction

India is known to possess a significant demographic dividend due to its large and growing population where the female population is estimated to be 586.5 million or above currently. This population is characterized by vast regional and cultural differences that create difference in terms of region and culture due to which any kind of generalization is not possible. Women play a crucial role in the growth of the economy and over the years, they had made a substantial impact and achieved status across different sectors both within the country and overseas. Women is a part of workforce but still there is a large population of women both in rural and urban areas that has dreams, but lack the opportunities to realize those dreams and take up their roles other than the household ones. In this regard, it is essential for to promote skill development and women empowerment. In order to upgrade employability and other skills among women, there is need to reduce the unemployment, demographic dividend, cultural and regional barriers and bridge the gap between talent and industry by various innovative methods.

Proactive involvement of society is essential to attain the sustainable development and bring the desired change in the society. Women in economically disadvantaged areas lack opportunities for employment and for self-employment due to various causes and sometimes women in need are on the look-out for affordable vocational courses that is not only of interest to them but also enables them to enhance their employability. Women from these areas can be equipped with the skills they need to increase their employability so that they can access meaningful and sustainable employment, and self-employment through entrepreneurship. UNICEF is also assisting women to build stronger links with the private sectors so that they have sustainable market access for their artisanal goods including those who are running home-based business.

India is one of the fastest growing economy and a farm-based country where more than 50% population engages in farming for their livelihood. The food processing is of utmost importance for development because of vital linkages between the two pillars i.e. our industry and economy. It is considered as the sunrise sector as every state has its own food cuisine and specialty and from the age of 10-14, most of the rural girls learn many things from their elders especially mothers and grandmothers, that we can compulsory introduce in food processing practice at the small level for them to introduce women and girls as food entrepreneurs. Food processing sector is witnessing a wave of change wherein new conducive policies are being framed up and huge investment is flowing in to make it a global food processing hub in the near future.

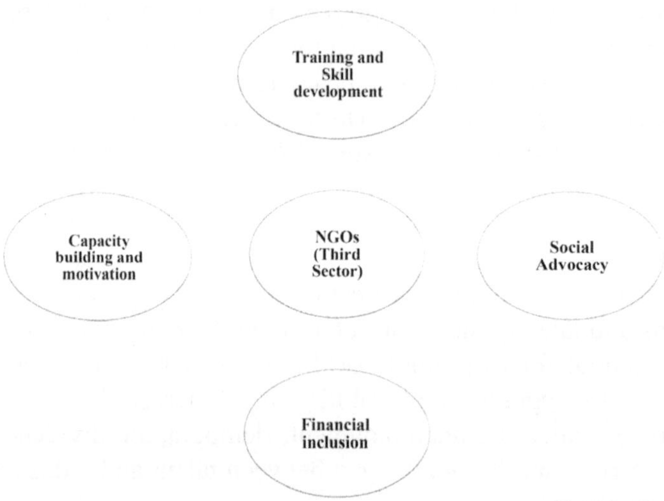

Fig. 1: Dimensions of NGO (Non-Governmental Organization)

killing and re-skilling of women is the need of hour and training them is an important aspect for women skill development and is also useful for income generation through self-empowerment. Training and upgrading the skills of in food industry sector is a key to improve the employability and build confidence in women and enhance sustainable rural development. They can be empowered with skills of numerous techniques of food preservation, processing and packaging as well as generates diverse opportunities which could be a great factor for their social upliftment. To ensure this and make this concept work, NGOs (Non-Governmental Organizations) are well positioned within this pool as they have developed capacities at the academic and practical level, as well as high level of acceptance at the grassroots level (as in Figure 1).

Current Scenario of Food Industry in India

Food processing sector holds a vast potential for growth as India has large production as well as consumer base. Over the past years, agricultural production has consistently recorded higher output and India has also ranked 1st in total food production globally. With respect to specific commodity, it ranked 1st in the production of milk, bananas, guava, papaya, ginger, okra, pulses and buffalo meat. Further, India has also ranked second in the world production of green peas, potatoes, tea and tomato. Our country produces almost all types of food due to our wide range of agro-climatic conditions. After all these productions, approximately 40 per cent of fruits and vegetables are wasted every year due to improper handling, transport, storage, marketing and post-harvest losses (Kumar *et al.*, 2017). These huge post harvest losses are of great concern as being a population of 1.3 billion where the 65 percent of them is below 35 years of age, with large disposable income and consumer preferences and especially women constituting 48.5 percent of total population can surely contribute their side of skill if given the opportunity.

The Indian Food industry accounts for 32 per cent of country's total food market and contributes 8.8 per cent of Gross value added (GVA) in manufacturing, 13 per cent of India's export and almost 6 per cent of industrial investment. India is not only a large producer of food but also has vast growing consumer base but despite its strong agricultural production base, a significant amount of food produce gets wasted in India due to inadequate infrastructure such as packaging facilities, storage, transport and low levels of processing.

It is clear that India has achieved a place in production of food but still opportunity in processing of various sub segments of food in country is still very low as compared to other developed nations. Only 10 per cent of agricultural produce is processed and rest 90 per cent remains unprocessed (DD News, 2017). According to Make in India, 2016, overall processing level of perishable products as compared to others is: India (approx 10%), USA (80%), Malaysia (80%), France (70%), Thailand (30%), Australia (25%). The variation in these numbers is due to the lack of training, capacity building and life skill to women for building their confidence and leadership. Further, due to the economic barriers and lack of infrastructure in food processing sector we are unable to use our production fully. With this large production, India now have to see a way forward for the opportunities and in spite of only dependent on government initiatives there are various NGOs also working in regional and rural areas at the grassroots level to make the food processing sector rise to it utmost height. Several unexplored opportunities underlying under the food processing sector is been mentioned in Table 1.

Table 1: Unexplored opportunities underlying in the food processing sector in our country

Segments (% of processing)	Primary processing	Secondary processing	Tertiary processing
Fruits and Vegetables (~2%)	Cleaning, cutting, sorting)	Pulp, Flakes, Paste, Frozen, Diced, Canned	Jams, Jellies, Chips, Ready to serve drinks, Indian ethnic drinks
Grains and Cereals	Sorting and grading	Rice puff, Flour, Baby food (Final product/ Ingredients)	Cakes, Biscuits, Breakfasts, Breads and other bakery products, RTC/ RTE products
Oilseeds	Sorting and grading	Cakes, Refined oils	Soya oil, Olive oil, Mustard oil, Fortified oil
Milk (Dairy ~35%)	Grading and refrigeration	Packaged milk, Flavoured milk, Cream, Milk powder	Yoghurt, Cheese, Ice-cream, Curd, Baby food, other value added products
Meat and Poultry (~21% and ~6%)	Sorting and refrigeration	Chilled frozen products	Ready to eat products
Marine products (~23%)	Chilled or Frozen products		Ready to eat products

(Source: Make in India, 2016)

It is clear from the Table 1 that we have various unexplored opportunities and it could be achieved by skilling and re-skilling of our women through NGOs involvement. NGOs can be used as an effective mechanism for exploring and disseminating the opportunities among the women and supporting them in socially and mutually learning, thereby achieving their sustainable and equitable participation.

New Dimensions of NGO Involvement

Non-Governmental Organizations (NGOs) are those organizations which may have an institutional base, or private, non-profit, self-governing, and voluntary in nature. They get financial assistance from individual donations, foreign funding and aid from different government agencies and private donors. There diverse work helps rid the prejudices, illiteracy and poverty in some context, thereby uplifting women of our country. Civil society has three functions to perform i.e. to inform, empower and ensure accountability. Whenever the government gets distracted, Non-governmental organizations are there for more peoples' participation and good internal governance for accountability and transparency. NGOs in India have spread like wildfire in the last decade and they are becoming the lifeline of the marginalized and disadvantaged communities, giving people a reason to live. NGOs are serving the people at different parts of the country and are led by an amazing and dedicated group of people resolving and working on major issues like women's empowerment and transforming the society.

Women constitute approximately half of the population and economic growth of the nation would not be balanced without the involvement of women in economic activities. There is wide disparity in women's economic status and there is need to empower and skill them building a bridge to eliminate this gap. To harness the entrepreneurial activities of Indian women and building capacity among them Non-Governmental Organizations (NGOs) can work at the desired level and strategic approach should be adopted. They are more critical and strategic in the process of global human development and acts as the third sector. For a decade now, Non-Governmental Organizations are actively involving and participating in combating challenges related to education, health, vocational training, women empowerment, skill development, etc. They take new challenges (Skilled workforce in Food processing sector) as well as opportunities (Unexplored opportunities in food processing sector) for fulfillment of their goal. As specific problems require a specific solution, there is no better place than NGO to understand the pulse at grassroots level and engage closely with the communities. They are working relentlessly on the ground providing critical services to the disadvantaged and vulnerable sections of the society. NGOs are known and trusted by the communities where they work, they have been helping people in their regional areas as well as other places to combat the stigma of fear to come forward.

NGOs played an important role to assist the community for improving their lives and situation where they use approaches (Top to down, Bottom to up and partnership approaches) for women empowerment and their skill enhancement. Participation and societal factors like motivation and social awareness is a central component and principle foundation of community development (As shown in Figure 2). NGO is a private organization that pursues activities to relief sufferings, provide essential social services, trainings and undertake community development as whole. The primary objective of NGO is to provide social justice, development and human rights but many of them are providing the additional services like training and capacity building programs for women. India has at least 31 lakh NGOs that means, its more than double the number of schools, 250 times the number of government hospitals. But these statistics doesn't correlate with the performance and result of developmental process. Figure 2 shows the pyramid of skill development from lower to higher level.

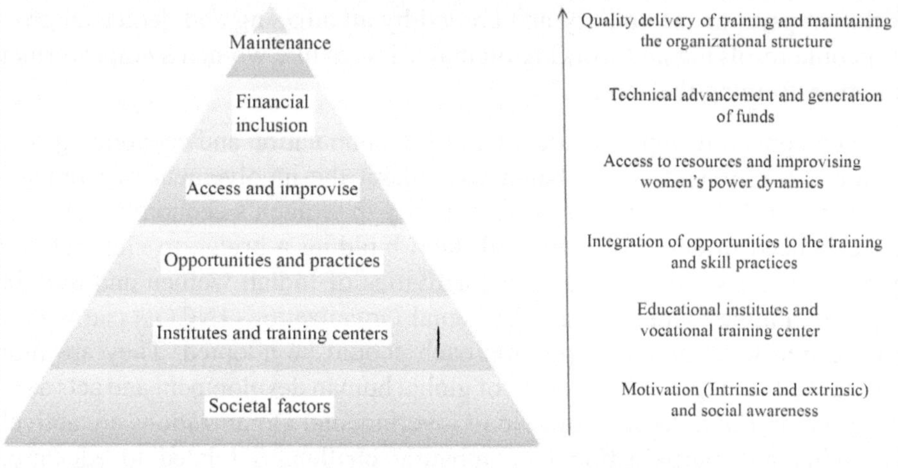

Source: As compiled by author

Fig. 2: Pyramid of skill enhancement (From lower to higher level)

Strategic framework for empowerment of women

Women have emerged as a diverse interest group in the last few decades due to the concept of equality. Women's economic participation and empowerment are fundamental to strengthening women's right and enabling them to have the control over their lives and exert influence in the society. Gender equality and empowered women are the catalysts for multiplying developmental efforts. Government of India has enacted various rules and regulations within the constitutional framework to improve female representation in different professions. Currently, majority of female workforce in India is unskilled and skill is the bridge between job and workforce.

Globalization and increased competition have intensified the need for highly skilled workforce in the economy to meet global standards of quality. Skill India and other initiatives seeks to give all Indians an opportunity to aspire and achieve better future for themselves as only 4.69 per cent of Indian population has undergone formal skill training. A combination of demographic, economic and social factors makes skill development an urgent priority for India. In this regard, a conceptual framework has been designed to use the indigenous knowledge of women regarding her regional foods and cuisines and skilling (and re-skilling) them with new practices so that there is desirable change in their behavior and they can finally become potential to use their skill for income generation, which ultimately empowers them (As in Figure 3).

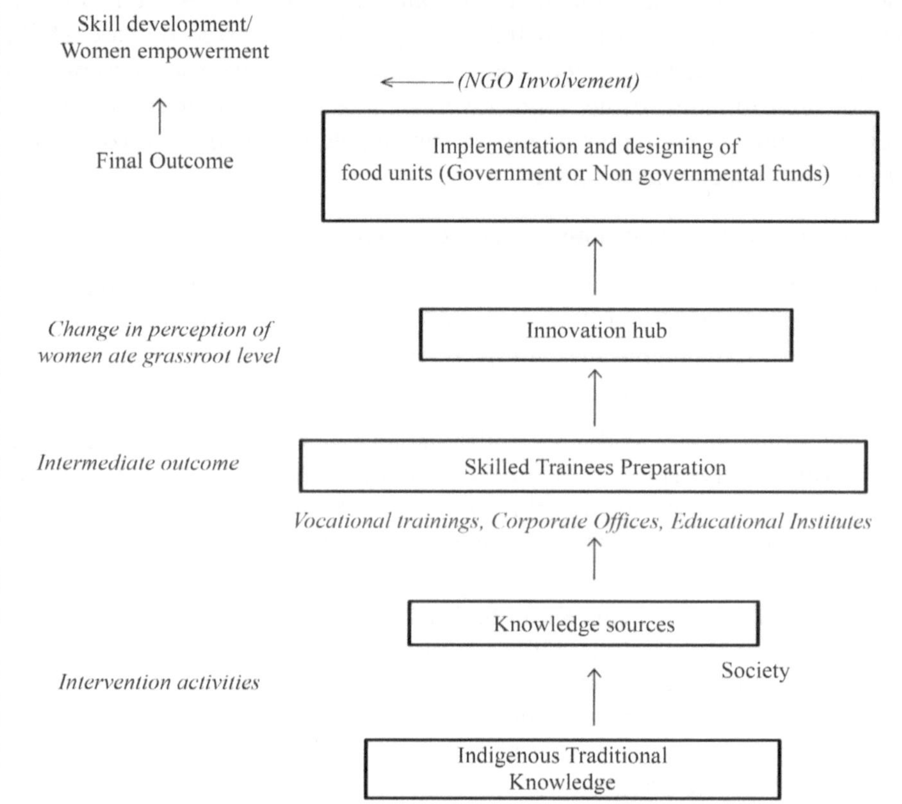

Source: As compiled by the author

Fig. 3: Conceptual framework of skill development through NGO involvement

Women's control over the income and decision-making is not that efficient and they don't have time for proper training in institutes. Women did not spend on food for household, but men did, potentially indicating greater control over income by men. Capacity building of women in food processing sector is need of an hour as they would assess the impact over several resources. Several vegetable production or home/ kitchen gardens along with the combination of animal husbandry can lead to the source of women's income and control over several resources.

NGOs are working effectively and efficiently in the field of women empowerment and their upliftment at the community based levels. NGOs are undergoing in training programmes related to food processing and other sectors and are also trained in techniques of fund raising and leadership skills. *Rural Litigation and Entitlement Kendra,* a leading NGO of Uttarakhand has been conferred with the future leadership award for its work in the field of women's empowerment. There were many small and medium cateogories of NGOs from all over the country at the end of an all India competition organized by Rockefellar foundation and EdelGivefoundation. This award was given to acknowledge the potential to achieve the good governance, effective programme impact, exemplary human resources practices, sustainable resource mobilization and transparent financial management.

Skill India Initiatives in Food Processing Sector

Availability of skilled manpower has been identified as one of the major challenges in Indian Food Processing industry and according to the Ministry of Food Processing Industries (MoFPI), there are various skill development initiatives taken by the government:

- MoFPI is closely working in collaboration with Food Industry Capacity and skill Initiative (FICSI), Sector Skill Council (SSC) in food processing and regularly guiding and assisting to achieve the mandate. There is conduction of regular meetings with stake holders along with the courses to review the progress.
- Under Pradhan Mantri Kisan SAMPADA Yojana, MoFPI has formulated a scheme for assisting skilling infrastructure and development of course curriculum. The scheme has two components: a) Development of course curriculum for training modules and its translation in English, Hindi and regional languages. b) Assistance for creation of infrastructure facilities for Skill Training Centres.
- FICSI is working on identification of job roles and competencies required for each job role so as to develop National Occupational Standards (NOS) for different sectors of food processing.
- NIFTEM and Indian Institute of Food Processing Technology (IIFT) are also conducting regular programmes/ courses on skill development and entrepreneurship for youth, women, farmers, self-help groups and industries. The Ministry of Food Processing Industries is very keen on entrepreneurship development among various groups through skilling/

capacity building, outreach programmes and village adoption programme by the two institutions under this ministry.

SAMPADA Yojana (Scheme for Agro-marine Processing and Development of Agro-processing Clusters) is a comprehensive package which will result in creation of modern infrastructure with efficient supply and will provide a big boost to the economy creating huge employment opportunities in rural areas, reducing wastage of agricultural produce, increasing the processing level, and enhancing the export of the processed foods. Figure 4 shows NGOs that are actively participating in community mobilization and women empowerment.

(Spiti)Ecosphere	Bakri Chhap
Ecosphere is a social enterprise which is a collaborative effort of the local community of Spiti and professionals from diverse backgrounds, with a wide spectrum of skills ad experience, effectively spanning the bridge from the general to the niche. **Focus area:** To achieve a responsive synergy in establishing the crucial link between economic empowerment, development and conservation through their range of products and services. Products available in their outlet: • Seabuckthorn product: Jams and concentrate • Organics: Black pea (Kala matar), Barley • Handicrafts, solar energy, etc.	Green people is a for-profit impact organization working towards integrated rural development via two channels namely "The Goat Village" and "Bakri Chhap". They are promoting farm retreats/ home stays and reviving abandoned villages by leveraging tourism potential and the latter by providing market linkage to indigenous farm produce from the marginalised farmers community that they work with. "Chote Kisano ka Bada Brand, Bakri Chhap " Multiple government and tourism organizations, luxury hotels, wellness experts, food critics are visiting the goat villages and buying their products. Products available: Preserves and beverages, lentils, breakfast cereals and millets, rice and flour, etc.

Source: As compiled by the author

Fig. 4: NGOs working in women empowerment

Constraints to women involvement in training and skill development

Women have emerged as a diverse interest group and despite their considerable involvement and contributions, significant gender inequalities exist in access to technologies, credit, information, inputs and services, probably because of inequalities in ownership of productive assets including their less involvement in decision-making and lack of technical guidance. Following are the constraints involved in training and skill development of women:

1. Inadequacy in collective action in community and people's mobilization
2. Lack of motivation and social awareness among the women

3. Inadequate knowledge regarding the financial inclusion and use of resources judiciously
4. Lack of popularity of skill and training amongst them
5. Untidiness created by semi-processing hinders in their participation
6. Lack of opportunities and skilled motivated personnel in their area of expertise

Strategies and interventions to unlock the potential of women in skill development

Skilling and re-skilling in rural India is really important to unlock women's potential in food processing sector. There are various strategies discussed below to unlock the potential of women in skill development:

1. There are opportunities available, but there is need to aware the women of our society about the skill initiatives in food processing sector to upgrade and improvise their skills and practices.
2. Information Communication Technology (ICT) and Mass media can be used in unlocking the potential of women by providing them trainings and courses that are available according to their suitability and compatibility.
3. There is immense potential of food processing and value addition, women are needed to search for a good product that is available in abundance in their own locality and regional areas, especially in niche areas such as organic and fortified foods and making them available at home itself.
4. Focusing on sweet revolution to boost India's export of honey. Presently, India ranks 6th in production and export in the world.

Conclusion

Unlocking the potential of women is the need of an hour and Non-governmental organizations are today, being explicitly used to women's empowerment in powerful ways as a source of funding and delivering training or employment opportunities for women as a way to create economic empowerment through micro-entrepreneurship and providing affordable services to women. Women contribute in the development process but they are still in grip of economic dependence and some of other latent factors inside them largely affect their behavior that is reflected as their acceptance and rejection. NGOs in India are promoting women participation in various programmes by addressing some issues concerning the capacity building and skill enhancement with extensive community participation. NGOs being a new dimension in today's world are developing the individual ability and performance of women by motivating and changing their behavior, awareness building, training and mobilizing women to participation. Thus, it can be concluded that rural development, skill enhancement and women empowerment is crucial for a developing country like India which has both producer and consumer base and NGOs are helping women community in self-reliance and micro-financing, and finally contributing towards the sustainable community development.

References

DD News. (2017). PM Narendra Modi address gathering at the Word Food India 2017. Retrieved from https://www.youtube.com/watch?v=3N0EUCcVo&t=637s on 28th November 2020

Kumar, N. B. Bhardwaj, S. Ahmad and Indu. (2017). A survey: To estimate food wastage periodically. International Journal of Engineering Research and Technology, 5(11): 1-4.

Make in India, Opportunities in Food processing sector. (2016). Retrived from https://www.yesbank.in/..../make_in_india-_opportunites_in_food_processing_sector.. on 28th November 2020

Pradhan Mantri Kisan SAMPADA Yojana (2017). Retrived from mofpi.nic.in/schemes/pradhan-mantri-kisan-samapada-yojana on 27thNovemeber 2020

Retrieved from https://mofpi.nic.in/investor-facilitation/skill-development-initiative on 29th November 2020

31

Integrated Agricultural Interventions for Improving the Socio-Economic Status of SC and ST Farmers under TSP and SCSP Programme

***Shubhadeep Roy*[1], *Neeraj Singh*[2], *Saikat Maji*[3], *Ankita Sarkar*[4] *and Kalyan Ghadei*[5]**

[1]*Associate Professor, I.Ag.Sc., BHU, Varanasi*
[2]*PrincipalScientist, ICAR-Indian Institute of Vegetable Research, Varanasi*
[3]*Assistant Professor, I.Ag.Sc., BHU, Varanasi*
[4]*Assistant Professor, I.Ag.Sc., BHU, Varanasi*
[5]*Professor& Head, Dep. of Ext. Edu.,I.Ag.Sc., BHU, Varanasi*

Scheduled Caste Sub Plan (SCSP) and Tribal Sub Plan (TSP) were initiated by the government to guarantee funds to SC/STs both at State and Centre in proportion to the size of their respective population. It was in 1974-75 in the Fifth Five Year Plan period that the Government of India introduced the policy of the Tribal Sub-Plan (TSP) and later in 1979-80 in the Sixth Five Year Plan period the Special Component plan (SCP) which was renamed as Scheduled Caste Sub Plan in April, 2006 and renamed as Allocation for the Welfare of Scheduled Caste (AWSC) in February, 2017.Specifically, the SCSP and TSP mandated that public resources towards SC/ ST welfare must be earmarked in proportion to their share in total population.The Narendra Jadhav Committee was constituted by the Planning Commission in 2010 in this regard.Every year, the Union Budget makes allocations exclusively for the Scheduled Caste and Scheduled Tribe.This fund is spent through the Scheduled Caste Sub Plan (SCSP) and the Tribal Sub Plan (TSP).As per the recommendations of the committee the earmarking of funds for all Central Ministries/Departments taken together should be at least 16.2% and 8.2% of the total Plan outlay respectivel under the SCSP and the TSP. The aim of such dedicated allocation is to promote nutritional security, livelihood security and entrepreneurship among these communities.

Stride of ICAR-IIVR and Operational Area

Amid nutritional and livelihood security of tribal population ICAR- Indian Institute of Vegetable Research (ICAR-IIVR), Varanasi has adopted 2724 tribal households from 24 villages of six clusters namely Faripan, Manvasa, Bakuliya, Chetwa, Jugail, Dahakudandi of Sonbhadra district of Uttar Pradesh and one cluster from Ranchi, Jharkhand for implementation of TSP programme of the institute. Similarly 2566 Scheduled Caste (SC) households have been adopted from 33 villages of five different clusters of Varanasi, Mirzapur and Chandauli districts for implementing SCSP programme.

Resource Specific Interventions for Profitability

The SC/ST farmers are basically resource poor and operates in economic threshold level. Their habitats are also environmentally disadvantageous. So the interventions done among these communities were according to their available resources and as per their working knowledge.The interventions were categorized infollowing modules that have been explained here.

1. Crop Based Module

A selected number of interventions were done in *crop based module.*Seed replacement for better yield, more nutrition and enhanced income was the motto of this module. Rice-wheat was the most prevalent cropping system in the villages adopted. However, it was realized that the system was not sustainable due to various reasons. The dominant rice – wheat cropping system of cultivation had led to serious economic, social and ecological problems such as deceleration in the productivity of ground water resources, decline in soil fertility and increase in environmental pollution. Need was felt to diversify the traditionally grown less remunerative cropping system through suitable interventions of vegetable crops to achieve sustainable productivity, profitability and resource use efficiency.Rice growers used to grow rice varieties of longer duration such as MTU-7029, Moti and Damini and other locally available cultivar. The use of long duration rice varieties invariably delayed the wheat sowing. Due to delayed sowing of wheat in December-January, the yield realized in wheat was often very poor. Under TSP and SCSP Programme rice growers were provided with certified seeds of medium duration (120-25 days), high yielding rice variety HUR-917 and HUR-3022developed from BHU which are aromatic in nature, resistant to lodging and neck blast disease with good grain quality. The improved high yielding medium duration rice varieties HUR-917 and HUR-3022 fetched higher yield than other varieties grown by farmers of similar growth duration. The timely sowing of wheat resulted in increase of yield which otherwise often get decreased due to delayed sowing. The delayed sowing of wheat resulted in availability of shorter growing period. Besides, early onset of high temperature during February, March often resulted in forced maturity and shrivelled grain in wheat leading to poor yield. The farmers were provided certified seeds of high yielding wheat variety, HD-2967, HUW-234 and also Zinc fortified wheat var. *WB-02* was introduced to combat Zn and Fe deficiency, particularly among women and children. The certified seeds of high yielding varieties improved the seed replacement rate in the area that resulted in higher yield due to low incidence of disease and pest infestation. Increased yield and income were witnessed due to intervention, in case of paddy 15.3% increase in yield and 22.5% increase in income and in case of wheat 28.9% increase in yield and 39.6% increase in income was recorded.Need based demonstration of pulses like arhar*var*. UPAS-120, Malviya chamatkar, NDR-2, til*var*. Sekhar, lentil *var*. HUL-57, gram *var*. Pusa-362, urd *var*. T-9, mung *var*. HUM-216 were conducted.

Fig. 1: Demonstration of high yielding paddy HUR 917 resulted 15.3% increase in yield as compared to the practicing cultivars in Sonbhadra district of Uttar Pradesh

2. Horticulture Based Module

In *horticulture-based module* of intervention, improved high yielding disease and pest resistant varieties of tomato, brinjal, cowpea, bottle gourd, sponge gourd, chilli, pumpkin, okra etc. were promoted in the form of kitchen gardening as well as commercial basis. Adoption of these varieties at large scale has led to increased productivity, production and profitability to the technology adopters. The adoption of these varieties by the farmers has affected the farming system in tribal areas by increasing cropping intensity by 191%.The vegetable area was only 1.34 % of the total cultivated area of the village. The major constrained for the area under vegetable was low yield and high cost of private company hybrid seeds, lack of knowledge of plant protection measures. The introduction of ICAR-IIVR varieties in the area and the increased knowledge and skill acquired by the farmers due to frequent training and exposure visit of farmers to the ICAR-IIVR Research Farm, resulted in more number of farmers taking up the vegetable cultivation as an enterprise and consequently the area of vegetable increased to 4.6 percent of the total cultivated area.

Table: Yield and income enhancement through vegetables

Crop (Variety)	**Average Yield (q/ha)**		**Yield Difference**	**Increase in Yield (%)**	**Increase in income (%)**
	FLD	**Control**			
Tomato (Kashi Aman)	390.2	280.3	109.9	39.21	25.02
Brinjal (Kashi Uttam, Kashi Taru, Kashi Sandesh)	403.7	325.5	78.2	24.02	27.09
Cowpea (Kashi Kanchan, Kashi Unnati)	115.3	91.7	23.6	25.74	79.78
Bottle gourd (Kashi Ganga)	270.1	230.5	39.6	17.18	46.05
Sponge gourd (Kashi Divya, Kashi Saumya)	95.4	77.9	17.5	22.46	56.70
Chilli (Kashi Anmol, Kashi Abha, Kashi Tej)	105.5	70.7	34.8	49.22	32.20
Pumpkin (Kashi Harit)	290.2	215.2	75.0	34.85	38.89
Okra (Kashi Pragati)	132.1	90.2	41.9	46.45	36.54
Pea (Kashi Udai, Kashi Naandini, Kashi Mukti)	80.3	55.5	24.8	44.68	39.91

Fig. 2: Vegetables cultivations in the kitchen garden for food and nutritional security

3. Livestock Based Module

Under *livestock-based module* backyard poultry was promoted and 10,000 number of one day-old-chicks of CARI- Debedra, CARI- Nirbheek and Kadaknath were procured from ICAR-CARI, Izatnagar and demonstrated among 500 landless beneficiaries along with starter feed and water feeder. Vaccination of the backyard poultry was conducted in the villages for sustainability of the model. Trainings were provided on poultry nutrition and hygiene and master trainer on poultry vaccination were prepared among the village youths. The backyard poultry provided an opportunity to earn extra income through selling of egg and bird. The introduction of improved breeds like CARI- Debedra, CARI- Nirbheek fetched better income to the farmers in shorter time with less investment and recurring contingencies leading to better benefit: cost ratio.

Fig. 3: SC/ST community people raring poultry in their backyard Impact of the interventions

Tribal sub plan was started from the year 2013 and SCSP is operational from the year 2019-20. Under these two flagship programmes of the institute interventions were carried out in the form of demonstrations of the need based technologies and capacity building activities of the stakeholders. Under TSP programme 15724 demonstrations were conducted in vegetable crops, 19615 demonstrations in field crops, 28062 demonstrations in nutria garden module and 22442 tribal beneficiaries were trained under different capacities to improve their livelihood. Similarly under SCSP programme 6416 demonstrations in vegetables, 5180 in field crops, 7535 in nutria garden module were conducted and 5093 SC beneficiaries were trained in different capacities for self sustainability. As a result of these efforts the average family income of the tribal beneficiaries has been increased to Rs. 133000 in the year 2021-22 as compared to Rs. 42000 in the base year 2013-14. Similarly average family income of SC beneficiaries has raised to Rs. 107500 as compared to Rs. 71500 in the base year 2019-20.

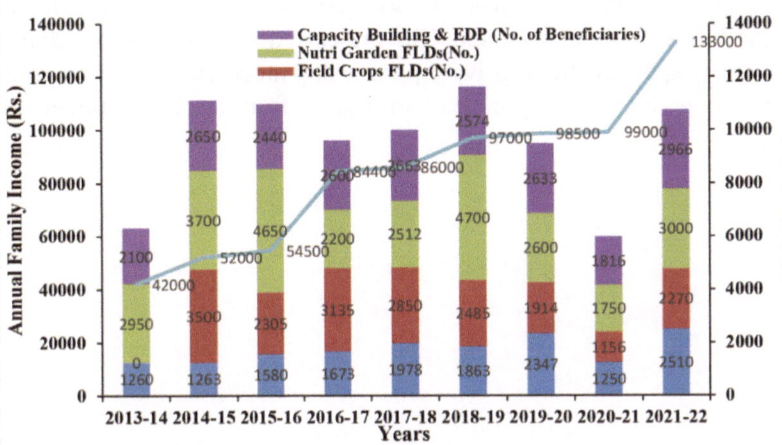

Fig. 4: Impact of intervention over period of time under TSP

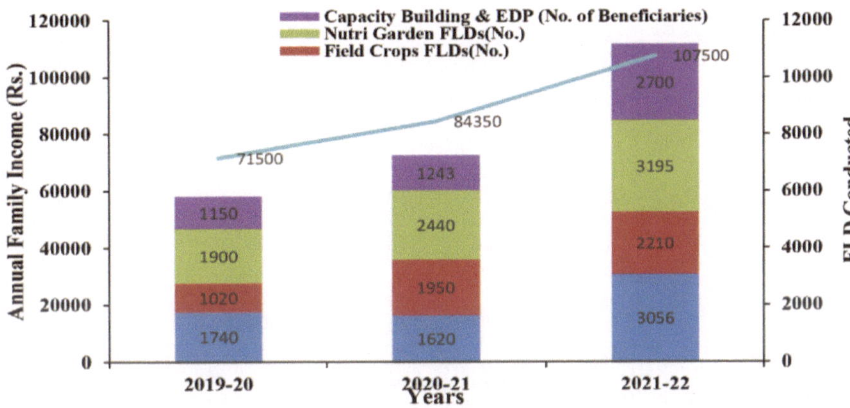

Fig. 5: Impact of intervention over period of time under SCSP

Conclusion

SCSP and TSP programme has become a ray of hope for the small and marginal farmers in the adopted villages. The unique concept of the project has directly benefitted the farmers which they did not experience earlier from any other extension agency. The farmers learnt to belive that they have strengths in their available resources and their working knowledge is equally important like many others. Now time has come to spread the benefit of the project to the nearby villages through Carpet extension or Horizontal extension approach. Further action is required to make all the intervention modules sustainable. Intervention required towards value addition of the raw materials like post harvest processing of the vegetables and egg products which has tremendous market value. Branding and lucrative packaging of the products is next target for capturing best market value.

Proper agriculture waste management and biogas generation in the project villages need to be done for healthy environment. The project villages may be developed as Agro-eco-tourism spots in public-private-partnership mode. Advanced trainings are required for achieving such objectives and support from the Government and policy makers.

Index